Fundamentals of Soil Science

A Laboratory Manual

Third Edition

Jeffory A. Hattey and Jaime J. Patton
Oklahoma State University

KENDALL/HUNT PUBLISHING COMPANY
4050 Westmark Drive Dubuque, Iowa 52002

ISBN 978-0-7575-5645-6

Printed in the United States of America
10 9 8 7 6 5 4 3

Fundamentals of Soil Science: A Laboratory Manual

Table of Contents

Preface

This laboratory manual was written for an introductory level soil science course and targeted toward those who have a professional interest in the areas of agriculture, engineering, environmental science, forestry, natural resources and wildlife. The goal of this laboratory manual is to open students' eyes to the wonderful world beneath their feet, introduce them to fundamental soil functions and processes, and stimulate thinking on how soils influence our everyday lives.

Many individuals will be involved in management decisions regarding land use in their professional and personal lives. These laboratory exercises are designed for students to make observations and measurements of soil properties they will frequently encounter. When direct measurements of soil properties or processes are not involved, exercises are designed to help students develop mathematical and reasoning skills that will be applicable in their future endeavors. There are many creative soil science instructors who have supplied ideas for some exercises adapted into this laboratory manual. They are acknowledged at the end of each exercise.

We thank the many students, who made constructive comments about the exercises, editing comments, and have worked through the exercises. We especially like to thank all the laboratory instructors (24 to date) who have put in a considerable amount of time and effort to help develop the laboratory exercises. Two students whose help was invaluable in developing several exercises are Mr. Rick Kochenower and Mr. Damon Wright. Finally, we thank our families who are always so supportive.

J.A. Hattey
J. J. Patton
June 2002

EXERCISE 1

Soil Characteristics and Formation

OBJECTIVES

1. *Define the soil*
2. *Identify the components of soils*
3. *Note differences that occur within and between soils*
4. *Learn how soils form*
5. *Observe arrangement of different properties in a soil profile*

INTRODUCTION

The **soil** is a **dynamic natural** body composed of **minerals, organic materials, air**, and **water** that **acts** on and is **acted** on by **living organisms** in a **thin layer** covering the earth's **surface.**

The four non-living components of soil (Table 1.1) are in active equilibrium with soil organisms to form a living body, a dynamic natural resource. Soils have many functions in our everyday lives. They act as a reservoir for nutrients and water, as well as physical support for plant growth. Soils also provide the base for the roads, highways, and building foundations that comprise a civilization's infrastructure. Soils are also important in the decomposition and recycling of waste products generated by animals and humans. This recycling process not only breaks down our trash, but reintroduces many elements important for plant growth back into the environment. In addition, much of the world's carbon is stored in the soil organic matter. Currently, soil carbon sequestration is being targeted as a potential sink for excess atmospheric CO_2, one of the primary gasses associated with global warming. Because of the importance of soils in our lives, we must learn to use this valuable natural resource in the most effective manner, not only for ourselves but for the generations to follow. Therefore, the foundation for our appreciation of soils begins with understanding the processes involved in soil formation.

Table 1.1. Soil components and approximate percentage of each

Component	Description	Approximate Percentage
mineral	weathered rocks and minerals	40-50
soil organic matter	organic material in various stages of decay	0.5-10
air	N_2, O_2, CO_2 other gases	25
water	solvent in soil	25
living organisms	bacteria, fungi, worms, protozoa	variable

Soil Genesis (Formation)

There are **five soil forming factors** that play a significant role in soil formation. They are :

1. **Biota**
2. **Climate**
3. **Parent material**
4. **Time**
5. **Topography**

These soil forming factors interact with each other in varying degrees during the process of **soil formation (genesis)** to form a countless array of soils. Of these five factors, climate has the most significant effect on soil formation. **Climate** influences the rate of soil formation, by controlling the weathering rate of minerals that form the material from which soil arises, known as soil **parent material**. Parent material in turn effects the soil's physical and chemical properties, including the size and reactivity of soil particles. **Biota** plays a major role in the organic portion of soil formation. Plants "fix" C from the atmosphere through the process of photosynthesis. When they die, the plants deposit organic matter into the soil, which is then subjected to decomposition by a myriad soil organisms. Soil organic C and the organisms that utilize it significantly affect many physical and chemical processes in soil formation. **Topography** affects the energy and erosional aspects of soil formation. In landscapes with steep slopes, the rate of soil erosion may be as great or greater than the rate of soil formation thereby limiting the depth of soil genesis. It is important to remember that it takes **time** for the soil forming factors to produce even a small amount of soil in this dynamic system.

The soil forming factors in a soil system work in concert with the **soil forming processes** of **addition, transformation, translocation,** and **loss** to produce the existing soil. Addition of organic matter to the upper portion of the parent material is significant factor of soil genesis. As organic matter accumulates, the surface layer takes on a dark-brown to black color as organisms transform the organic matter. Organic acids formed as a result of this transformation decompose minerals that in turn alter the nutrient, water, air and status of the soil and result in the promotion of plant growth. Soil constituents are transported through the soil by water, air, and organisms resulting in different zones within in soil. Zones of **eluviation** (leaching) and **illuviation** (influx) result in horizontal layers containing different chemical and physical properties called soil **horizons.** These zones are never static as new materials are constantly added, transformed, translocated or lost from the system. Losses from the system are in the form of gases released from microbial respiration or evaporation constituents in solution through leaching, runoff, or solids in the form of soil erosion or plant removal. The result of the additions, transformations, translocations and losses is a distinctive set of unique soil horizons.

A **soil profile** is a collection of horizons that are arranged in a unique fashion due to variations in the soil forming factors and processes. An upper horizon with a significant accumulation of organic matter in various states of decay is called an **O horizon**. O horizons are most prominent in soils that form under forests or in cold or wet regions where organic matter decomposition is limited. Not all soils will contain an O horizon. Below the O horizon (if there is one), is typically the **A horizon,** which is the upper mineral horizon. The A horizon contains an accumulation of partially decomposed organic matter (humus). Unlike the O horizon which is dominated by organics, the A horizon is dominated by minerals. The **E horizon** is a zone of leaching or **eluviation**. In the E horizon, organic matter, clays, Fe, Mg, and plant nutrients have been leached out leaving only SiO_2 (quartz) and other materials highly resistant to weathering. The E horizon is typically very light in color due to the lack of clays and organic materials and the dominance of light colored silicates. Like the O horizon, E horizons are typically found in soils that form under forest vegetation. The **B horizon** is a zone of accumulation or **illuviation** (washed in). In the B horizon, there is typically an accumulation of clays, carbonates, Al- and Fe-oxides, or salts. The B horizon is lighter in color than the A horizon because less organic matter is present, but darker than an E horizon due to the accumulation of clays or oxides. The **C horizon** lies below the B horizon and is composed of parent material that has not been subjected to soil forming processes. The **R horizon** underlies the other soil horizons and is composed of consolidated bedrock. When you look at a soil profile you will notice some obvious differences. They are:

- **Color**
- **Texture**
- **Structure**

Soil Color

As we look at a soil profile, color is the most obvious property that differs by horizon and by soil profile. Color itself has little impact on soil physical and chemical properties, but is rather an indirect indicator of current physical and chemical properties or a sign of past soil conditions. Dark brown to black colors near the soil surface usually indicate the presence of organic matter. Reddish colors indicate the soil contains iron oxide minerals. A region in the soil profile that contains mottles of reddish-yellow mixed with blue-gray colors (redoximorphic features) indicates that water is (or has been) present at this level during some periods of the year. In this instance the red-yellow soil colors indicate periods of oxidation (O_2 present), while blue-gray colors indicate periods of reduced conditions (no O_2). **In order to facilitate communication between soil scientists, soil colors are described using a standardized format.**

Munsell Color Notation

Soil colors are determined using the Munsell color notation system. This system contains a collection of charts that include color chips of set **hue, value** and **chromas**. For color identification of soils, only a portion of the Munsell system is needed. Charts that contain hues of 10R, 2.5YR, 5YR, 7.5YR, 10YR, 2.5Y, and 5Y, are adequate for most soils. Two special charts that contain gleyed colors are used for soils in reduced (wetland) environments.

Hue

Hue is the dominant or spectral color of the rainbow. There are 10 primary hues (Figure 1.1.a), and these hues are further divided into quarters (Figure 1.1.b). Each of these quarters (e.g. 2.5YR) is the basis of a page in the Munsell soil color charts. Increasing numbers within a hue indicate the increasing dominance of the prominent color in the hue. For example, as the color wheel moves from 2.5YR to 10YR, the amount of yellow contained in the hue increases. Each page in the soil color book contains one-quarter hue with a series of color chips that are categorized by value and chroma.

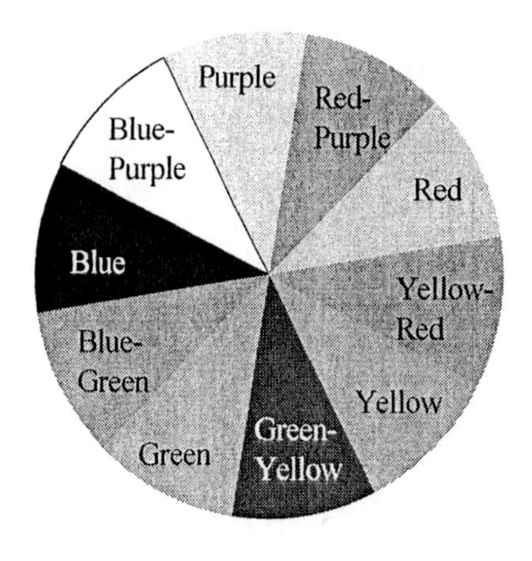

Figure 1.1.a. Hues used for classification in the Munsell color notation.

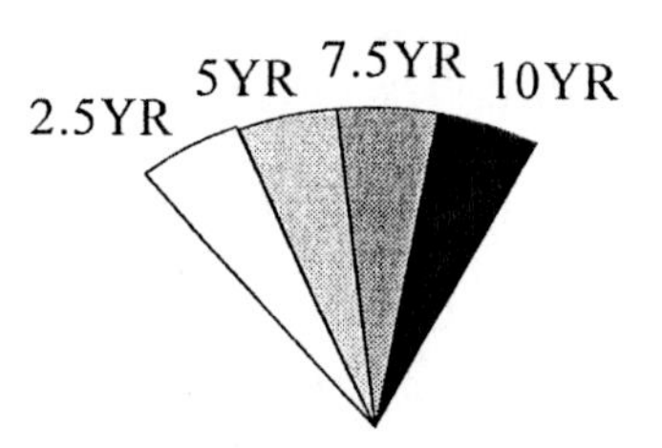

Figure 1.1.b. Sub-divisions of the Yellow-Red hue. More Yellow is added as the scale moves from 2.5 to 10YR.

Value

Value is the amount of white or brightness in a color. If one were to make a diagram of values, it would go from black at a value of zero to white at a value of 10 (Figure 1.2). The values for soil colors range from 2.5 to 8 and form the rows within a Munsell color chart.

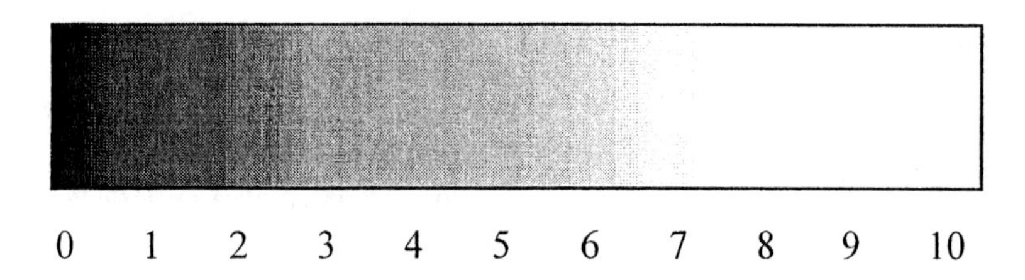

Figure 1.2. The range for value as it moves from 0 to 10.

Chroma

Chroma is the amount of pigment that is present in a color. For example, chroma would be the quantity of 2.5YR pigment added to a bucket of white paint to get a desired color of reddish-yellow to paint a house. Figure 1.3 shows how increasing amounts of black pigment are added with each chroma increment for a gray value of 5. At a chroma of 1, there is very little black pigment within the color and so the color is seen as white. As the chroma increases, additional black pigment is added and the color changes from white, to various shades of gray, and ultimately black. Chroma forms the columns in the Munsell color chart.

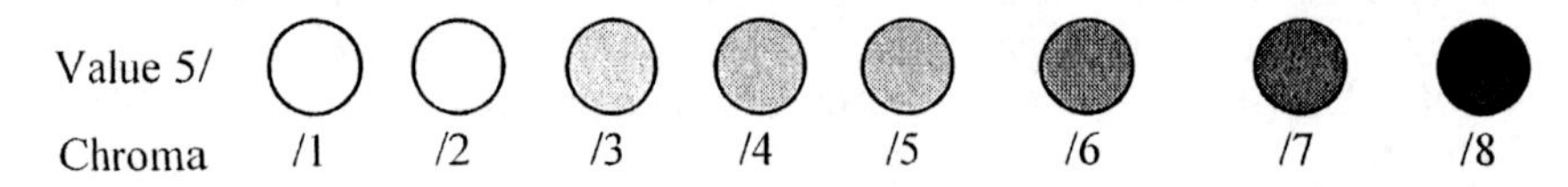

Figure 1.3. Chroma changes for a gray value of 5.

To properly describe a soil color, one must use the color's hue, value and chroma. An example of proper color notation for the Ap horizon of the Dennis series is10YR 4/2, where hue = 10YR, value = 4, and chroma = 2.

Soil Texture

Soil texture is the relative distribution of the three **soil separates; sand, silt**, and **clay**. The separates are classified by their size, which is discussed in more detail in Exercise 6. For now, understand that sand particles are larger than silt particles, which are larger than clay particles. The texture of a soil is a constant property relative to our lifetime. Soil texture can only be modified by major forces such as mineral weathering, geologic events such as floods or erosion, the addition of wind deposits of silty or sandy material, or soil additions by man. Soil texture influences characteristic soil properties such as: how soil particles arrange themselves to form aggregates, infiltration rates, water storage capacity, plant available water, and the type and quantity of nutrients available for plant growth.

Initial characterization of soil texture can be done by feel. When moistened, soil can be squeezed between the fingers to estimate the proportion of sand, silt and clay in the sample. The estimated proportions of the separates are used to categorize the soil texture based on USDA soil classifications. Examples of USDA textural classifications include: loam, silt loam, and sandy clay to name a few.

Soil Structure

Soil structure is influenced by the interaction of the soil separates and soil organic matter to form soil **aggregates** (Figure 1.4). An individual aggregate is called a **ped**. Peds vary widely in size (from approximately the size of a BB to bowling ball), but all are held together by the attractive forces associated with clay particles and soil organic matter. Peds have a tendency to break along fracture planes when subjected to an external force, in a manner similar to a mineral crystal, to form repeating shapes and sizes. Peds should not be confused with clods, which are formed by soil disturbance and unnatural fracturing or with concretions, which are high concentrations of a compound that are cemented together. Soil structure is categorized by the shape and size of the ped.

Ped Structures Found In Soils

Granular peds are rounded, sphere-like aggregates that lack sharp natural cleavage planes. Granular peds look similar to marbles and are typically small in size.

Platy peds are plate-like and flat, with a vertical dimension much shorter than that of the other two dimensions. Platy structure looks similar to parent material derived from shale and is often found in the E horizon or in areas of compaction.

Blocky peds are cube or block-like in shape with all three dimensions being approximately the same size. When the soil is broken apart, blocky peds have sharp natural cleavage planes. There are two sub-types of blocky structure:

angular blocky: flat faces and very sharp angles
sub-angular blocky: slightly rounded faces and angles.

Prismatic peds are rectangular shaped with the vertical dimension several times greater than the horizontal dimension. The vertices of prismatic peds are angular and the peds have flat top and bottom surfaces.

Columnar peds are similar to the prismatic peds except the top of the structure is rounded. This structure is associated with soils containing high levels of sodium.

There are also two categories of non-structure:

Single grain soils have no aggregation and are non-coherent. Single grained, structureless soils is similar to sand on a beach, in that the individual grains act independently of each other. .

Massive soils have no observable aggregation and are coherent. They have no observable shape pattern or cleavage planes when an external force is applied

Granular structure is typically found in the A horizon where there are significant levels of organic matter. Soils containing granular structure usually have high amounts of available water for crops, as well as good aeration due to the large number of pores present. Blocky, prismatic, and columnar structures are normally found in the B horizon, where the highest quantities of clay are generally found. Single grain and massive structures are generally associated with the C horizon, where the soil forming processes have been limited and little aggregation has occurred.

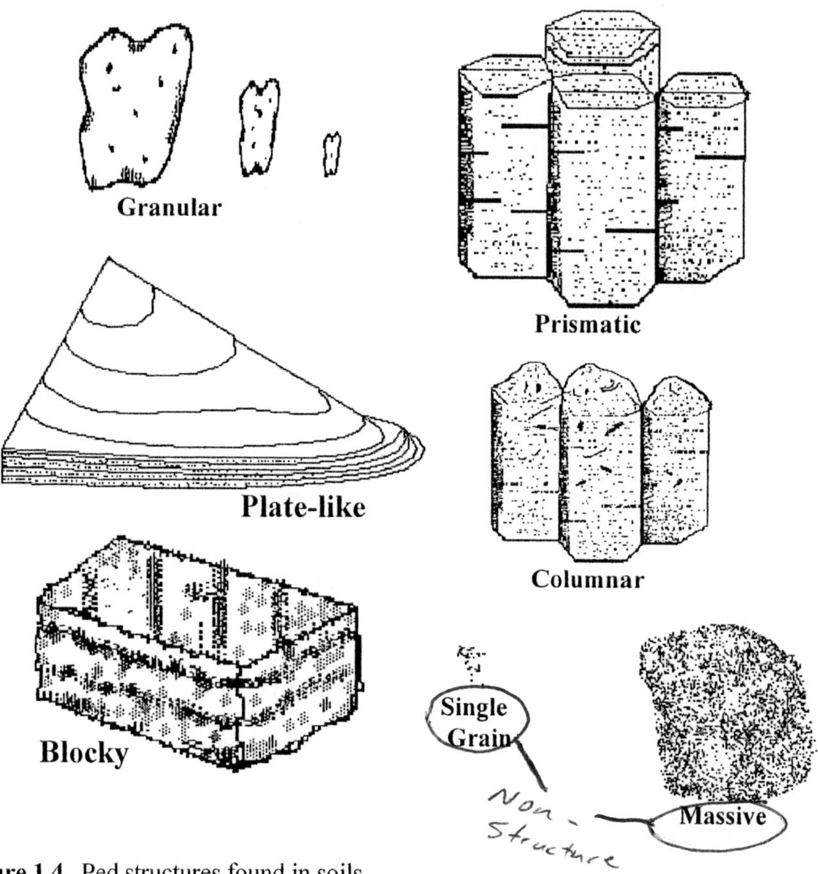

Figure 1.4. Ped structures found in soils.

Review Questions

1. What is the soil?
2. What are the five soil forming factors?
3. What does color indicate about a soil?

4. What are the important processes for soil genesis?
5. What is the difference between soil texture and soil structure?

References

Beyrouty, C.A., N. Slayton and M.G. Hanson. 1992. Lab Manual for Soils. University of Arkansas, Fayetteville, AR.

Weil, R.R., 1981. A Laboratory Manual for General Soils. Avery Publishing Group Inc. Wayne, N.J.

DEMONSTRATIONS

Soil Color

Two samples have been set up to practice determining soil color. Use the Munsell color chart to determine the hue, value and chroma for each sample. To accurately use the color chart, the soil should be moist (not saturated) and full light should be used. To do this, select a soil ped then:

1. Moisten slightly using a water bottle
2. Hold the sample behind a page in the color charts so it can be seen though the hole in the page. Match the sample with the color chip closest to the soil's true color. Record the hue, value, and chroma in Table 1.2 designated for soil color.
3. Experiment with matching the sample with color chips on different pages. Observe the differences in colors as you switch between hues, values, chromas.

Caution: Do not get the color chips wet or dirty. This will ruin an expensive book!

Soil Texture

Three samples (sand, silt, clay) have been set-up to feel soil texture. Place a small sample in the palm of your hand. Add water to moisten the sample, and then work it between your thumb and forefinger. Note the different feel of each texture. One will feel gritty, another smooth, and the third will feel smooth and sticky. Can you determine which is which? Record your observations in Table 1.3.

Soil Structure

There are several peds set-up for you to observe soil structure. Take measurements of the dimensions of the peds. Where would you expect to find each of the structures: in the surface soil or the sub-surface horizon? Write your answers in Table 1.4.

Soil Profiles

There are four soil monoliths laid out on the lab benches. Chose any two monoliths and take the following measurements or observations:

1. Depth of each horizon
2. Soil color for each horizon
3. Structure for each horizon

and record in Table 1.5.

PLEASE TAKE THE FOLLOWING PRECAUTIONS:

1. **DO NOT WET THE MONOLITHS TO DO COLOR!**
2. **DO NOT REMOVE ANY SOIL FROM THE MONOLITHS TO DO TEXTURE OR STRUCTURE!**

Table 2.2

EXERCISE 2

Minerals and Rocks Important to Soils

OBJECTIVES

1. ***Review the minerals and rocks which are important for the formation of soils***
2. ***Identify selected minerals and rocks***
3. ***Identify physical properties of minerals and rocks that affect soil characteristics***
4. ***Identify chemical properties of minerals and rocks that affect soil characteristics***

ORIGIN OF SOIL MATERIALS

Soils can be divided into two categories: mineral soils, which are composed primarily of weathered minerals and rocks, and organic soils, which have a significant amount of decaying plant and animal materials. Most soils are mineral soils, although organic soils are very important and productive in some areas of the world. For example, the organic soils of Ireland, Minnesota, and Florida are used extensively for vegetable production, as well as peat mining. The characteristics and productivity of a soil are often related to the chemical composition of the minerals and rocks from which they originate. Soils that develop from sandstone rocks will tend to be sandy, well drained soils with low water holding capacity. Soils that develop from shales tend to be clayey, poorly drained soils with a high water holding capacity. Soils which develop from rocks and minerals high in Ca, K, Mg tend to be very productive agriculturally, whereas soil developed from rocks and minerals high in Al, Si, and Fe are often more acidic and contain fewer plant nutrients. These innate properties of the mineral derived soils will in turn affect many other physical and chemical properties, as we will see in later exercises.

Minerals

Minerals are defined as "... a naturally occurring homogeneous solid with a definite (but not generally fixed) chemical composition and an ordered atomic arrangement" (SSSA, 1997). Elements that comprise the minerals in the lithosphere (earth's crust), hydrosphere and upper 10 miles of the atmosphere are shown in Table 2.1. Many minerals in the earth's crust are composed of Si, Al, Fe, and O. These elements combine to primarily form silica, aluminum and iron oxides, which comprise 74.7% (SiO_2=59.8% and Al_2O_3=14.9%) of the upper portion of the earth's surface. Silica, iron, and aluminum oxides typically weather very slowly with minerals containing silicon weathering the slowest. Therefore, minerals that contain a high percentage of silica such as chert and quartz, are very resistant to weathering and often comprise a large percentage of rocks in humid regions. Other minerals containing calcium, phosphorus, potassium and other elements necessary for plant growth are also found in the earth's crust.

There are two categories of minerals: **primary minerals** which are unaltered materials formed from cooling and solidification of molten materials and **secondary minerals** which are altered materials formed from the decomposition and resolidification of primary minerals (Table 2.2). Primary minerals form in high energy environments (i.e. intense pressure and/or heat), whereas secondary minerals form in low energy environments (i.e. conditions at the earth's surface). Since primary minerals contain more high energy bonds than secondary minerals, they are generally less stable at the earth's surface, and thus subject to faster rates of weathering. Both primary and secondary minerals significantly influence the chemical and physical properties of soils that are derived from them.

Rocks

Rocks are composed of two or more minerals and can be placed into categories of igneous, sedimentary, and metamorphic.

Igneous rocks comprise 95% of the material in the earth's upper crust (Table 2.3). They are formed by the cooling and solidification of molten materials at or below the earth's surface.

Table 2.1. Elemental composition of the upper 10-mi. of the earth's crust

Element	Lithosphere	Hydrosphere	Average (Including Atmosphere)
Oxygen	46.60	85.79	50.02
Silicon	27.72		25.80
Aluminum	8.13		7.30
Iron	5.00		4.18
Calcium	3.63	0.05	3.22
Magnesium	2.09	0.14	2.08
Sodium	2.83	1.14	2.36
Potassium	2.59	0.04	2.28
Hydrogen	0.22	10.67	0.95
Titanium	0.46		0.43
Carbon	0.19	0.002	0.18
Chlorine	0.06	2.07	0.20
Bromine		0.008	
Phosphorus	0.12		0.11
Sulfur	0.12	0.09	0.11
Barium	0.08		0.08
Manganese	0.08		0.08
Nitrogen			0.10
Fluorine	0.10		0.10
Others	0.50		0.47
Total	**100.00**	**100.00**	**100.00**

Table 2.2. Chemical composition and characteristics of selected primary and secondary minerals that are important in soil formation

Name	Chemical Composition	Classification	Color
orthoclase	$KAlSi_3O_8$	primary	White to red
labradorite	$(Ca,Na)Al_2Si_2O_8$	primary	Dark to greenish gray
hornblende	$(Ca, Mg, Fe)(SiO_3)_2$	primary	Black
biotite	$H_2K_2(Mg, Fe)_2Al_2(SiO_4)_3$	primary	Black
olivine	$(Mg, Fe)_2SiO_4$	primary	Green
talc	$3MgO{\cdot}4SiO_2{\cdot}H_2O$	primary	White
kaolinite	$Al_2O_3{\cdot}2SiO_2{\cdot}2H_2O$	secondary	White
bentonite	$H_6Al_2Si_4O_{14}$	secondary	White-gray
calcite	$CaCO_3$	secondary	Colorless
dolomite	$CaMg(CO_3)_2$	secondary	White gray
quartz	SiO_2	primary	Colorless white
chert	SiO_2	primary	Gray
hematite	Fe_2O_3	secondary	Red to black
gibbsite	$Al_2O_3{\cdot}2H_2O$	secondary	White
pyrite	FeS_2	secondary	Yellow brassy
halite	$NaCl$	secondary	Colorless white
apatite	$Ca_5F(PO_4)_3$	primary	Green

Table 2.3. Chemical composition of igneous and metamorphic rocks

Name	Type	SiO_2	Al_2O_3	Fe_2O_3	FeO	K_2O	CaO	Na_2O	MgO
		%							
Granite	igneous	70.9	16.2	1.6		5.3	2.9	1.3	
Gabbro	igneous	47.8	18.6	11.6		0.5	9.4	3.6	4.2
Basalt	igneous	49.1	15.7	5.4	6.4	1.5	9.0	3.1	6.2
Obsidian	igneous	76.8	12.1	0.6	0.8	4.9	0.6	3.8	0.1
Volcanic ash	igneous	>76.8	similar to obsidian except higher SiO_2						
Gneiss	metamorphic	74.7	8.9	9.6		9.0	1.1	0.4	1.9
Hornblende-schist	metamorphic	50.3	14.1	7.1	5.3	2.3	8.1	4.0	7.2

The chemical composition of igneous rocks is dependent on the elements contained within the molten material from which they formed and their texture is dependent on the rate at which the material cooled. In general, the slower the cooling process, the larger the crystal, and the coarser the resulting soil texture. Igneous rocks are further divided into categories based on their silica content. Acidic igneous rocks (> 65% SiO_2) are composed primarily of aluminosilicate materials and are considered light in weight and color compared to other rocks. Basic igneous rocks (< 50 % SiO_2) are composed of aluminosilicate materials, as well as Fe, Mn, Ca, Mg and Na bearing minerals. They are often considered heavy in weight and dark in color in comparison to other rocks. Igneous rocks between these ranges of SiO_2 content are considered neutral.

Sedimentary rocks are formed by the transport, cementation, and consolidation of weathered rocks, minerals, plant and animal remains or by the evaporation of mineral rich waters. **Clastic** sedimentary rocks are formed by cementation of minerals on the macro- to microscopic scale and are categorized by particle size and the type of cementation. Particles can be cemented together by calcium carbonate ($CaCO_3$), clay, organic matter, iron oxides, soluble aluminum or other materials. **Precipitate** sedimentary rocks are formed by the chemical precipitation of dissolved materials. The most common precipitate rocks are the limestones, formed from $CaCO_3$ deposition (Table 2.4). Sedimentary rocks generally form near the earth's surface in parallel beds ranging in thickness from a few millimeters to several meters.

Table 2.4. Selected sedimentary rocks and the characteristics used to categorize them

Name	Type	Categorized by	Characteristics
Conglomerate	clastic	size > 2.0 mm	rounded gravel and stones
Breccia	clastic	size > 2.0 mm	angular gravel and stones
Sandstone	clastic	size 2.0 to 0.05 mm	cemented sand grains
Shale	clastic	size < 0.05 mm	often horizontal layering
Marl	clastic		freshwater shells
Coquina	clastic	precipitate of $CaCO_3$ and shells	
Limestone	precipitate	precipitate of $CaCO_3$	>55% $CaCO_3$
Magnesium limestone	precipitate	precipitate of Mg,$CaCO_3$	
Oolitic Limestone	precipitate	precipitate of $CaCO_3$	Small, rounded nodules
Gypsum	precipitate	$CaSO_4$	soft white
Halite	precipitate	NaCl	table salt

Metamorphic rocks are igneous or sedimentary rocks that have undergone mineralogical, chemical or structural transformations due to the application of heat, pressure, heat and pressure, thermal gases, or thermal solutions (Table 2.5). Although metamorphic rocks contain the same general chemical composition as their igneous or sedimentary counterparts, their physical properties are quite different.

Table 2.5. Original material of metamorphic rocks important for soil formation

Initial Material	Consolidated Materials	Metamorphic rock
granite		gneiss
gabbro		schist
basalt		schist
gravel	conglomerate	quartzite
sand	sandstone	quartzite
mud or clay	shale	slate
calcareous ooze	limestone	marble

Transformations of Minerals and Rocks

While it is informative to know about minerals and rocks, as soil scientists we are most interested in them after they have undergone the weathering process. The weathering process of minerals and rocks occurs at varying rates by both chemical and physical means.

Disintegration Physical processes resulting in the resizing of materials by:

1. Expansion and contraction due to temperature changes
2. Abrasion by wind, water, and ice
3. Physical activity of plants, animals, and humans

Decomposition Chemical processes resulting in the changes in chemical composition of the materials by:

1. Hydrolysis $KAlSiO_3O_8 + H_2O \leftrightarrow HAlSiO_3O_8 + K^+ + OH^-$
2. Hydration $Al_2O_3 + 3H_2O \leftrightarrow Al_2O_3{\cdot}3H_2O$
3. Acidification $2NH_4^+ + 4O_2 \leftrightarrow 2NO_3^- + 4H^+ + 2H_2O$
4. Oxidation/Reduction $4FeO + O_2 + 2H_2O \leftrightarrow 4FeOOH + 4e^-$
5. Dissolution $KCl \leftrightarrow K^+ + Cl^-$

As minerals and rocks weather they become the **parent material** for future soils.

In general, rock and mineral materials with basic elements (e.g. Ca, Mg, Mn and Fe) tend to weather faster than materials with more acidic elements (e.g. Al and Si). Thus, minerals such as quartz and chert are relatively resistant to weathering, and minerals such as calcite and halite weather very quickly. As soils form from weathered materials, they tend to retain properties similar to the rocks or minerals from which they were derived. Soils from basic materials are often more fertile for plant growth than soils derived from acidic materials. Soils derived from large grained/textured materials are usually well drained (i.e. sandy) and soils derived from fine textured/grained materials are often poorly drained (i.e. clayey). The infinite possible combinations of parent materials and the other soil forming factors result in a diverse population of soils.

Review Questions

1. What is a mineral?
2. What is a primary mineral?
3. What is a secondary mineral?
4. What four elements are most prevalent in the lithosphere?
5. What primary mineral is the most resistant to decomposition in the soil?
6. What is the definition of a rock?
7. What is the origin of igneous rocks?
8. How are sedimentary rocks formed?
9. What processes are needed for the formation of metamorphic rocks?

References

Jacobs, H.S. and R.M. Reed. 1964. Soils Laboratory Exercise Source Book. American Society of Agronomy. Madison, WI.

Reed, R.M., 1986. Fundamentals of Soil Science Laboratory Manual. Oklahoma State University, Stillwater, OK.

Soil Science Society of America. 1997. Glossary of Soil Science Terms. Soil Science Society of America. Madison, WI.

Procedures

1. Review the rocks and minerals provided on the lab benches. You should become familiar with the characteristics (e.g. mode of deposition, color, texture, and possible plant nutrients) of the samples provided.
2. Answer the questions on the hand-in sheet concerning selected minerals and rocks.
3. Fill in Table 2.6 on the various characteristics of the selected rocks and minerals. What type of crystal or grain does the original material have? Is it large or small? With this information what type of texture (e.g. coarse, medium, or fine) do you think the resulting soil will have? **Remember** that the texture is a function of crystal size AND chemical composition.

EXERCISE 3

Soil Sampling and Soil Profiles in the Field

OBJECTIVES

1. ***Learn how to take a soil sample for fertilizer (nutrient) recommendations***
2. ***Learn how to take a soil sample for environmental analysis***
3. ***To describe the horizons of a soil under field conditions***

INTRODUCTION

An important factor in proper management of soil resources is knowing what physical and chemical properties are present. These properties can be identified in a variety of ways. Soil information can be gathered from observing the soil profile, reviewing the soil survey manuals (Exercise 4), and by collecting soil samples. Your particular need for and use of soil information will determine which is the proper method of gathering soil data. In this exercise we will examine two types of soil sampling—agricultural and contaminant sampling and try our hand at describing soil profiles in the field.

Soil Sampling

For practical reasons we often obtain soil samples to assess characteristics of a soil. When taking a soil sample it is important to evaluate the area being sampled, as well as the type of analysis required. In agriculture, soil sampling and soil testing serves as a guide for the application of fertilizers for crop nutrient needs. If we are interested in soil contamination, soil samples can help us identify what chemicals are present in the soil and where they are located.

Soil Nutrient Status

To maintain optimum plant growth, the nutrient levels of a soil should be assessed on a regular basis by soil testing. It is best collect soil samples when the crop is not growing. Specifics for obtaining a representative soil sample for nutrient testing are outlined by Johnson and Allen in OSU Extension Facts, F-2207.

Sample Size and Area

When agricultural soil sampling, it is important to take a representative soil sample from the area of interest. Researchers have shown that a minimum of 20 subsamples taken at random from a uniform area are necessary to get a representative sample. Once 20 subsamples have been taken and placed in a plastic bucket, they should be mixed thoroughly, and a composite sample of this field is sent to a laboratory for analysis.

When sampling, it is important to consider the area being sampled, because the collected sample will be an average for the entire area. Large areas that have changes in soil texture, slope, organic matter, soil depth, or cropping scheme may need to be separated into different sampling areas and have their own soil samples sent to the lab.

Sampling Depth

Agricultural soil samples can either be taken using a sampling tool (available at your local county Extension offices), spade or shovel. Sampling depth is usually at the depth of tillage or where there is the greatest crop root activity. For most crops, sampling depth is usually assumed to be 15 cm (6 inches). This depth is also where most fertilizer and agricultural chemical applications are considered most effective. Crops that have a rooting depth greater than 15 cm (6 in.) may need to have an additional subsoil sample

(15 to 120 cm or 6 to 24 in.) collected. The subsoil sample is useful for determining the amount of additional mobile nutrients, such as nitrate-nitrogen (NO_3-N), available to deep-rooted crops during the growing season.

Contaminated Soils

Employees of environmental consulting firms or government agencies are often responsible for collecting soil samples on contaminated sites. Sampling contaminated sites is very different than sampling for soil nutrients. When sampling for contaminants, it is important to consider the site from a three dimensional aspect and to be aware of the chemical and physical characteristics of the possible contaminates.

In contaminant sampling, it is important to consider the third dimension (depth), so that the total volume of soil affected can be estimated. This is done by collecting a soil core to a desired depth and dividing the core into sections (Figure 3.1), so that the location and quantity of the contaminant in the soil profile can be determined. Each section of the soil core is treated as an individual sample and is **never** mixed with other samples.

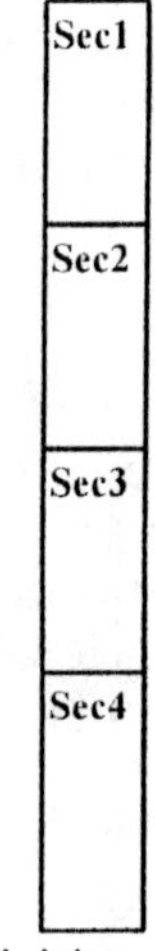

Figure 3.1. Division of a soil core for the purpose of contaminate analysis.

In addition to sampling dimensions, several other pieces of collection information are often needed to meet state or Environmental Protection Agency (EPA) **quality assurance/quality control (QA/QC)** requirements. A partial list of suggested soil parameters for site description are listed in Table 3.1. A QA/QC plan must be filed with the appropriate government agencies prior to sampling a site. This plan will often include information about how the site will be characterized, how the samples will be obtained, and once obtained how the samples will be handled prior to analysis. Once a QA/QC plan has been approved, it must be strictly followed or documentation must filed and approved for sampling alterations, in order for the samples to be admissible in court or other legal proceedings. Required equipment and documentation are available from the EPA (www.epa.gov).

Table 3.1. Suggested soil parameters to be determined for site evaluation.

Parameter	Soil Profiles	Soil Samples
Horizons	R	R
USDA Texture	R	R
Color	R	R
Redoximorphic features	R	
Porosity	R	r
Zones of Increased Porosity/Permeability		
Soil structure	R	r
Extrastructural cracks	r	r
Roots	R	r
Surface fractures	R	
Sedimentary features	R	
Zones of Reduced Porosity/Permeability		
Genetic horizons	R	
Consistency	R	
Root restriction layers	R	
Compaction	r	
Other		
Moisture condition	R	R
Water table	R	R
Saturated hydraulic Conductivity	R	r
Clay minerals	r	
Other minerals	r	
Odor	r	

R = recommended for all situations
r = recommended where climatic, geologic, and soil conditions make parameter significant
Adapted from: Boulding, 1994

Review Questions

1. How would you sample a field for a soil test if there was a saline slick 50 ft. in diameter in a 40 ha (100 A) field?
2. What problem(s) would arise as a result of collecting a soil sample from one location when sampling the field above?
3. Why do we take a 0-6 in. (0-15 cm) sample for soil testing purposes?
4. What are some different types of soil sampling tools?

References

Boulding, R. 1994. Description and Sampling of Contaminated Soil. 2nd. ed. Lewis Publ. Boca Raton, FL.

Procedure:

A. Taking a Good Soil Sample (Soil Testing)

We will practice obtaining good soil samples to be used for nutrient recommendations. The procedure is outlined as follows:

1. Visually separate the field into uniform areas from which a single, representative sample can be taken. Separation should be based upon the differences in topography, slope, soil texture, organic matter, and soil depth.
2. In Figure 3.2 delineate how you would separate the field for sampling purposes. Make a note of any regions to avoid while sampling.
3. Take a soil probe and push it in the ground approximately 15 cm (6 in.). Withdraw the probe and place in the collected core in a plastic bucket.
4. Collect 20 random cores throughout the area delineated for sampling, as illustrated in OSU Fact Sheet F-2207.
5. Once all 20 samples are placed in the bucket, mix them thoroughly and remove a subsample for analysis.
6. Place the subsample in a completed soil sampling bag and bring the sample back to the lab. **We will keep your sample for use in other labs later in the semester.

B. Soil Sampling Contaminated Soils

Rather than collect samples for contaminated soils, we will focus on describing the soil profiles as a means of site evaluation. This will be similar to Exercise 1, as we will be evaluating depth of soil horizons, and characteristics of those horizons.

Your laboratory section will be divided into three groups. At each site, the groups will have different responsibilities.

Group A:
1. Depth of master horizons in the soil profile
2. Identify the parent material

Group B:
1. Identify the soil color
2. Identify the soil structure

Group C:
1. Estimate topography (e.g. slope, upland, bottomland, flood plain etc...)
2. Diagram the soil profile (**Note** any unusual characteristics)

How to Get a Good Soil Sample

Oklahoma Cooperative Extension Service • Division of Agricultural Sciences and Natural Resources

F-2207

Hailin Zhang
Director, Soil, Water and Forage Analytical laboratory

Gordon Johnson
Nutrient Management Specialist

Soil tests provide a scientific basis for evaluating available plant nutrients in cropland, pastures, lawns, and gardens. Analyses of soil samples can help farmers and homeowners fine-tune nutrient applications from fertilizers, biosolids, and animal manure. Properly managing the amount of nutrients added to the soil can save money and protect the environment.

Soil nutrients vary by location, slope, soil depth, soil texture, organic matter content, and past management practices, so getting a good soil sample stands out as a major factor affecting the accuracy and usefulness of soil testing. This fact sheet outlines some specific considerations which should be taken into account to get the greatest benefit from soil testing.

Sample Soil at the Right Time

Fields used for production of cultivated crops may be sampled any time after harvest or before planting. Generally, two weeks should be allowed for mailing, analysis, and reporting of results. Additional time may need to be allotted for ordering and application of fertilizers, manure, or lime materials. Noncultivated fields should be sampled during the dormant season. In either case, do not sample immediately after lime, fertilizer, or manure applications because those samples do not represent the true soil fertility.

Fields should be tested annually to measure the available nitrogen pool or as frequently as necessary to gain an understanding of how soil properties may be changing in relation to cultural practices and crop production.

Collect a Representative Sample

Getting a representative sample is simple, but not easy. Research at OSU and other universities has clearly shown that a minimum of 20 cores or small samples taken randomly from the field or area of interest are necessary to obtain a sample which will represent an average of the soil in the field (Figure 1). These cores should be collected in a clean plastic bucket (to avoid metal contamination) and mixed thoroughly by hand. The sample bag should be filled from the mixture. A one pint (OSU soil sample bag full) sample is usually adequate for all tests which might be required. If the sample is too wet to mix, it should be spread out to dry some and then mixed, or sampling should be delayed until the field is drier.

It is important to remember that the sample obtained by the above procedure will be an *average* of the area sampled. If the area sampled is extremely variable in the soil properties which are going to be tested, then it may be better to separate the field into smaller areas, and get a representative (20 cores) sample from each of these areas in order to determine how variable the field is (Figure 2). In this way, it may be possible to treat some areas of the field differently from others and remove variability so that the field can be sampled and treated as a unit in the future. Variability in a field can often be noted by differences in surface soil color and crop growth or yield.

Using only one sample for a large variable field can be very

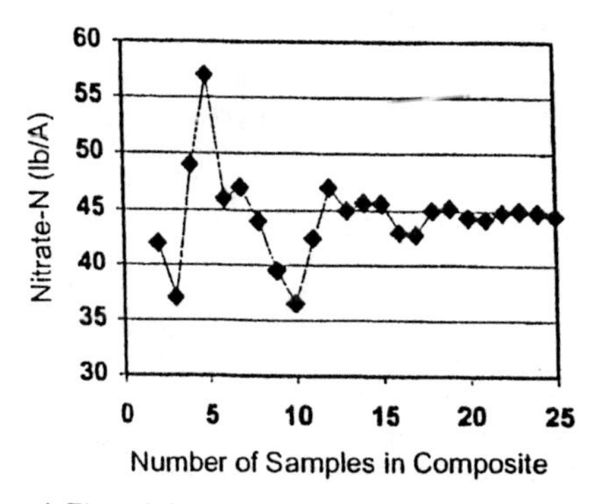

Figure 1. The minimum number of core samples needed to make a representative composite sample is about 20.

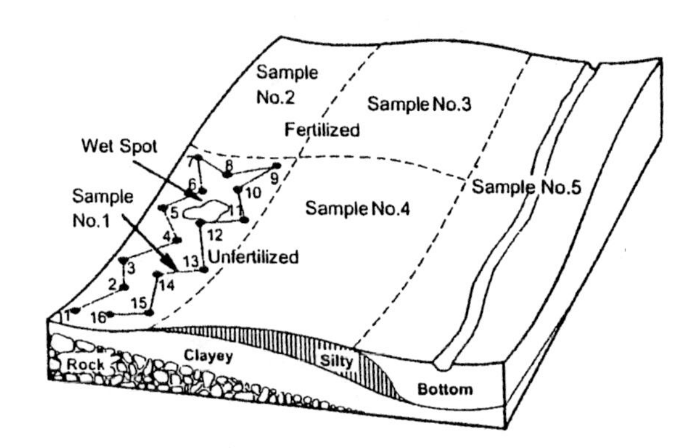

Figure 2. Divide field into uniform sampling areas and follow a random pattern when sampling. Avoid unusual spots and try to obtain a representative sample.

costly. Since the sample represents an average of the soil in that field, recommendations based on the soil test will likely cause the field to be overfertilized on some parts and underfertilized on other parts. Failure to obtain uniform response to treatments based on a soil test is frequently a result of one sample being used to represent a large variable field.

An example of field variability is shown in Table 1. The range of test values was obtained by testing 40 individual cores taken at random from an "apparently uniform " 80-acre field. The variation is great enough so that for some analyses the average is not a good representation of the field. Areas of the field with the lowest pH, phosphorus, and potassium values will not receive adequate lime or fertilizer if recommendations are based on the average test values.

A single core sample, or spadeful, is extremely risky because it may test anywhere in the range shown for each of the analyses. For example, deficiencies for wheat could range from zero to 37 pounds of P_2O_5 and zero to 34 pounds of K_2O. For alfalfa, which has much greater nutrient requirements, deficiencies could range from zero to 94 pounds of P_2O_5 and zero to 120 pounds of K_2O. This would also affect the amount of nitrogen and lime required. Obviously, unless the 80 acres is divided into less variable units for testing, some areas of the field will receive either too much or too little fertilizer and lime.

In deciding how large an area can be represented by one composite sample (20 cores), the determining factor is not the number of acres involved, but rather, the variability of the area. Some large, uniform fields can be represented well by a single 20-core sample, while some highly variable fields need to be split into two or more smaller areas for testing. Regardless of the field size or main area being sampled, unusual spots in the field (salty or wet spots) should be avoided during the initial random sampling. When unusual spots make up a significant area, they should be sampled separately.

Sample at Proper Depth

Cultivated Fields

For most soil tests the sampling depth is the tillage depth. The reason for this is because most crops have their greatest root activity in the tillage depth. Obtaining a representative sample with regard to depth means that each of the 20 cores taken from an area should be from similar depth, tillage, or six inches. Soil tests are generally calibrated on the basis of an acre furrow slice, approximately two million pounds of soil in the top six inches.

For deep-rooted nonlegumes such as wheat, bermudagrass, sorghum, and cotton, a separate sample representative of the subsoil should be taken in addition to the tillage depth or six-inch sample. This subsoil sample should represent the layer of soil from 7 to 24 inches below the surface. Because nitrate-nitrogen is mobile in the soil, a test of available nitrogen (and/or chloride and sulfate) in the subsoil sample will provide a more complete picture of available mobile nutrients for these crops (Figure 3) and can save fertilizer expenses.

Table 1. Variability of an 80 Acre Field Based on Soil Tests of 40 Individual Soil Cores .

	Soil Test Values	
Analysis	*Range*	*Average*
pH	4.9-6.3	5.6
BufferIndex	7.1-7.4	7.3
Nitrogen	1-34	11
Phosphorus	23-114	36
Potassium	149-770	306

No-till Fields

Noncultivated fields should be sampled to a depth of six inches, again because this is the effective depth of most treatments and the depth of most root activity. Nutrients from fertilizer, animal manure, and lime can be accumulated on the surface if they are surface applied without incorporation. A set of samples from the top two inches will help identify stratification of nutrients and is especially important for pH determination for no-till fields. If nutrient loss in runoff is the main concern, the two-inch sample is better than a six-inch sample because only the surface inch or two is in direct contact with surface runoff.

Salinity Diagnosis

When salt accumulation is suspected as a cause of poor stand establishment and the sample is being taken after planting, then the depth of sampling should approximate the seeding depth (one to three inches). This is especially important when conditions have been favorable for soluble salts to move upward and accumulate near the surface after planting. Since excess salts are most harmful to germination and seedling vigor, it is this shallow depth which should be tested. At other times during the year, a sample of the entire tillage depth may be most useful to test for salt accumulation.

Send Samples for Analysis

Soil sample bags are available at local county Extension offices. Extension offices will mail your samples to the OSU Soil, Water and Forage Analytical Laboratory and assist you to interpret test results.

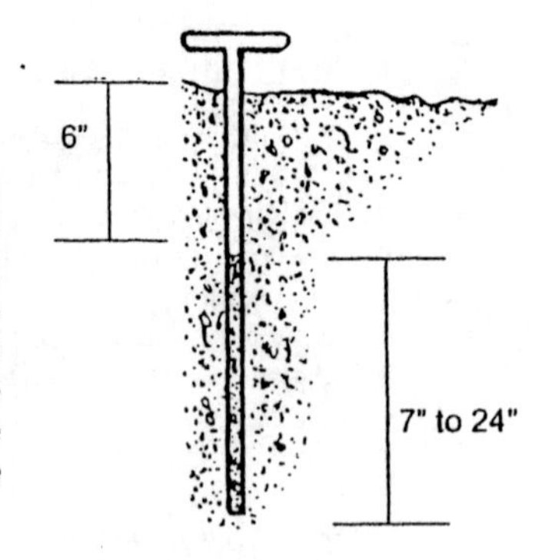

Figure 3. A soil probe is a good tool for obtaining soil samples. Push the tube to the six-inch depth and remove the core. Then take the seven to 24-inch core through the same hole for the subsoil test.

Issued in furtherance of Cooperative Extension work, acts of May 8 and June 30, 1914, in cooperation with the U.S. Department of Agriculture, Samuel E. Curl, Director of Cooperative Extension Service, Oklahoma State University, Stillwater, Oklahoma. This publication is printed and issued by Oklahoma State University as authorized by the Dean of the Division of Agricultural Sciences and Natural Resources and has been prepared and distributed at a cost of $000.00 for 2,500 copies. #3335 0399 CC Revised.

EXERCISE 3

Soil Sampling and Soil Profiles in the Field
Hand-In

Name:________________________

Lab Sec.________

Points__________

Questions:

1. How deep should a sample be collected from when soil sampling for K levels?

____

2. How many cores should be collected from a uniform 20 ha (50 A) field for fertilizer analysis?

____

3. How would your sampling strategy change if you want to document benzene contamination from an underground storage tank at the above site?

____
____
____
____

4. Why is it important to obtain a representative soil sample when nutrient sampling? What negative effects could you suffer if you collected a "bad" sample?

____
____
____
____

5. After collection, how would you treat a soil core from a contaminated site different than you would a soil core for a nutrient test? (Hint: Think of the major difference in sampling purpose).

____
____
____
____
____
____

Soil Profiles

Soil Diagram

Soil A:______________________________

Topography:__________________________

Parent Material:______________________

	Depth (cm)	**Color**	**Structure**	**Texture**
A Horizon				
B Horizon				
C Horizon				

Soil Diagram

Soil B:______________________________

Topography:__________________________

Parent Material:______________________

	Depth (cm)	**Color**	**Structure**	**Texture**
A Horizon				
B Horizon				
C Horizon				

Figure 3.2. Field diagram to delineate site for nutrient sampling

Field Diagram

EXERCISE 4

Soil Survey and Legal Descriptions

OBJECTIVES

1. ***To locate information about soil properties and management capabilities using a soil survey***
2. ***To learn how read and understand land legal descriptions***

INTRODUCTION

Soil Survey

The infinite combination of soil forming factors and processes create a diverse population of soils, so diverse that there are currently greater than 18,000 soil series recognized in the United States alone. While differing soil forming factors produce a large number of soil series, at the local scale sites that have the same soil forming factors will typically generate soils of the same series (with some slight variation). By recognizing the repeatability of soil forming processes on a small scale, we can begin to identify locations where individual soil series will occur and thereby develop a system to identify and delineate soil units in a given area. In the early 1900's, individuals with tremendous insight realized that soil information and land use could be combined with maps (e.g. aerial photographs) to determine where a particular soil could be found, thus the **Soil Survey** concept was born. Currently, three types of soil surveys have been produced: **Reconnaissance, Soil Conservation,** and **Standard Soil Surveys.** For our purposes, we are going to concentrate on the standard soil surveys.

Standard soil surveys are typically generated at the county level within each state. These surveys are compiled in a cooperative effort between the federal, state and county governments called the **National Cooperative Soil Survey.** At the federal level, the **Natural Resources Conservation Service (NRCS)** has the primary responsibility for producing the survey. At the state level, land grant universities have a significant input on data collection and analysis. Counties are often required to help fund the cost of producing the surveys. Surveys have been completed for most counties in the US. A list of completed soil surveys can be found at the National Soil Survey Center's website (http://soils.usda.gov). Print copies of a published soil survey can be obtained from one of the following sources:

1. Local Natural Resources Conservation office
2. County extension office
3. Congressional representative

Standard soil surveys are directed towards the needs of the population within the county and their use of the soil resources. Agricultural producers (farmers, ranchers, horticulturists) are often interested in how to use and manage soils, particularly when deciding where to locate row crops, forages, and trees and where soil conservation practices are needed. For engineering purposes, soil information is needed for highway, watershed, or conservation structure planning, as well as foundation and earthwork. In urban settings, soil information is needed for zoning, loans, tax assessment, sewage disposal, and recreation sites. Soil surveys contain information useful to all these users of soil resources.

Standard soil surveys are divided into two parts. The first part contains a written report about the nature of the county and descriptions of the typical soil profile for each soil series in the county. The survey also contains information on the genesis of each soil series as well abuse and management information. The written report is extremely useful for agricultural producers, engineers, and government planners as they decide how to optimize use of their soil resources. The second part of the survey consists of maps that show the location of each soil series in the county. **Soil surveys are valuable tools that are available to all individuals through the**

NRCS. In the future, as you make decisions for your personal and business ventures, be sure and take advantage of the plethora of information available in soil surveys.

Legal Land Descriptions

Every tract of land must have a legal description in order to pin-point its exact location. Two systems are commonly used in the United States, the **metes and bounds** system and the **rectangular survey**.

When the United States was first settled, the metes and bounds system was used in the eastern US and Spanish territories of the southwestern US to delineate parcels of land. The metes and bounds system uses a designated starting location from which distance and direction are given to the next corner. The description continues from corner to corner until the description returns to the starting point, thus describing the boundaries of the land. Metes and bounds descriptions often include prominent landmarks (trees, rocks and roads etc.) and are often irregularly shaped. Due to the frequent loss of identifying landmarks and complications in redeliniating complex boundaries, boundary disputes are common in areas where the metes and bounds system is used.

The rectangular survey system was first adopted in 1875, in what is now Ohio and this system is currently used in 30 states, including Oklahoma. The system uses **baselines** (east-west lines) and **meridians** (north-south lines) as reference points from which tracts of land are described. There are 34 principal meridians and one base line currently used in the US. In some states like Oklahoma, there are two principal meridians from which legal descriptions are based. In Oklahoma these principal meridians are the **Indian** and **Cimarron** meridians. The Indian Meridian is located in the central part of the state and is used for land descriptions in all counties located outside the Panhandle. The Cimarron Meridian is located at the western edge of Cimarron County and is only used in land descriptions in the Panhandle of Oklahoma (Cimarron, Texas and Beaver Counties). Therefore to eliminate confusion, all legal descriptions in Oklahoma have the notation of (C.M. or I.M.) at the end to designate their reference meridian.

The base unit of a tract of land is the Congressional **township** (not to be confused with the civil township, e.g. Stillwater, OK). Townships begin where a principal meridian and a baseline meet. They are laid out in 36 square mile blocks and are approximately 6 miles by 6 miles, except on the north side of a township where the boundary is about 50-ft shorter than the southern side due the curvature of the earth. Each township contains 36 **section**(s), with each section being one square mile (again as near as possible) or 640 **acres**. Parcels of land smaller than 640 acres are described as a portion of a section (i.e. quarter section). More details on the rectangular survey system are outlined by Kletke (OSU Extension Fact Sheet No. 9407).

The practical aspect of knowing about legal descriptions is to determine the location and area of a tract of land you are interested in.

REVIEW QUESTIONS

1. What organizations are responsible for producing soil surveys?
2. What are possible problems associated with the metes and bounds system?
3. How many principle meridians are in Oklahoma?
4. What information in a soil survey would be of interest for planning a city park?
5. If you were purchasing land for cotton production, how would a soil survey be of use?

PROCEDURE

A. Using a Soil Survey Manual

1. Use the soil survey manual to answer the questions in the hand-in.
2. Be sure to notice the amount of information for agricultural engineering, recreational, urban and wildlife uses of soil that is present.

B. Using Legal Descriptions

1. Read the fact sheet by Kletke and answer the questions in the hand-in.

Legal Land Descriptions in Oklahoma

Oklahoma Cooperative Extension Service • Division of Agricultural Sciences and Natural Resources

F-9407

Darrel Kletke
Professor, Agricultural Economics

"Hello, is this Mr. Ezra Jones? Mr. Jones I saw the article in the City Press on the superior breed of cattle you sell. How can I reach your place?

"Two miles south or a little more from Loyal and then 2 1/2 or 3 miles west? Where is Loyal? Northwest of Kingfisher? About 20 miles? Can you give me the legal description of the quarter on which your residence is located? Yes, I'll wait until you look it up.

"The NW quarter of Section 13, township 17 North, Range 9 west of the Indian Meridian? Thank you Mr. Jones, I can find you now because you've told me exactly where you live."

A legal description is to a farm what a street address and city is to an urban residence. Both show the exact location of a property, and each will be exclusive to that particular piece of realty.

In spite of the importance of legal descriptions in transactions involving real estate, many people have only a limited knowledge of what a proper description is, how to read or write it, or how it is determined.

The system used in most parts of the United States for describing the boundaries of rural ownerships is the envy of the world and because of its simplicity and it is easily learned by anybody. This publication will explain how land location is described in Oklahoma. The same system is used in all states west of the Mississippi, except in Texas, and in many states east of the Mississippi.[1]

The Rectangular Survey System

The shortcomings of indiscriminant settlement, the overlapping of claims, and boundary litigations were well known to the founding fathers. When new territory was being opened to the west, a Congressional Committee was charged in 1784 with the task of preparing a survey ordinance which would prevent a recurrence of boundary problems.

1 Many of the older states - the 13 original states plus Maine, Kentucky and Tennessee as well as Texas had considerable areas already settled when the system was devised. To avoid confusion of boundaries, the metes and bounds system was retained in some areas. In other areas a rectangular system was used, but these do not necessarily correspond with the system used by the U.S. Land office.

It recommended that all public lands be divided in "hundreds" of ten geographical miles square. The recommendation was tabled, but the following year the ordinance was amended to make the original "hundreds" six miles square and was passed by Congress in 1785. The rectangular survey system is confined to the public domain originally belonging to the Federal Government. Because of this, the states that originated from the public domain are often referred to as public-domain states.

The rectangular survey system is based upon the establishment of a principal meridian and a base line. The principal meridian runs in a true north and south direction; the base line runs east and west at a right angle to the meridian. The point where these lines cross is referred to as the initial point, or starting point. The geographic location of a meridian and base line is not fixed by law, therefore the 34 principal meridians in the United States were located to meet the convenience of government surveyors.

Land in Oklahoma is located with respect to two principal meridians; the Indian Meridian which runs from the Red River to the Kansas line through the central portion of the state, and the Cimarron Meridian, which runs the width of the Panhandle at its extreme western end. The base line to the Indian Meridian runs east and west from border to border across the southern part of the State. The base line to the Cimarron Meridian runs the length of the Panhandle along the southern border (Figure 1). All legal descriptions in Oklahoma, therefore, will end with the notation Cimarron Meridian (C.M.) or Indian Meridian (I.M.).

Townships. The legal description of a tract of rural land essentially is based on an area referred to as a Congressional township, and is not to be confused with a named Civil township. The Congressional township is an area six miles square and contains 36 square miles or sections.

After the principal meridian and base lines are laid out, the surveyor comes back to the beginning point and lays out lines north and south every six miles on either side of the meridian. He then lays out lines parallel to the base line six miles apart north and south of the base.[2] The lines surveyed result in a grid as illustrated in Figure 2. Each square in the grid is a township. The township marked "A" is referred to as

2 The actual surveying is a bit more complicated than this, but this describes the final results of the survey.

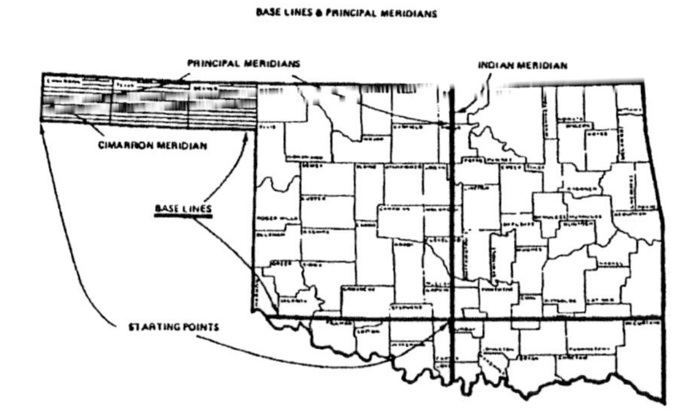

Figure 1. All legal land descriptions in Oklahoma include Indian or Cimarron meridian.

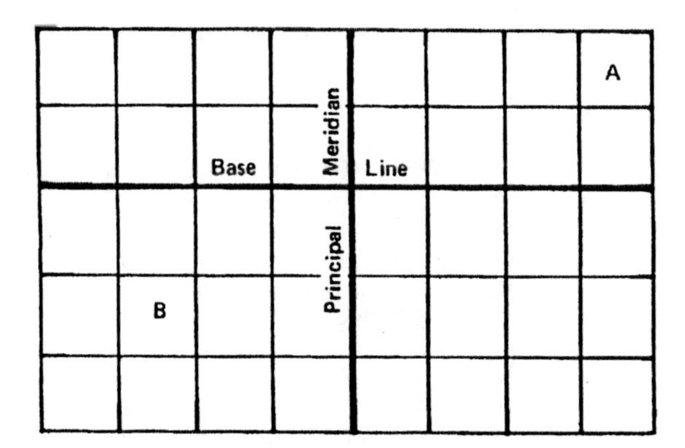

Figure 2. Squares six miles long on each side form townships on land maps.

6	5	4	3	2	1
7	8	9	10	11	12
18	17	16	15	14	13
19	20	21	22	23	24
30	29	28	27	26	25
31	32	33	34	35	36

N

Figure 3. Each township is further divided into 36 one-mile-square sections of land.

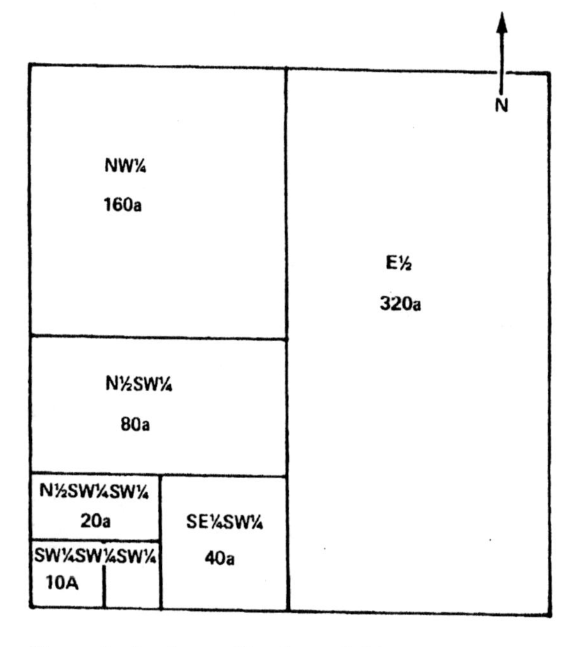

Figure 4. Sections of land are divided into quarters or smaller tracts as shown.

Township 2 North Range 4 East, because it lies in the second row of townships north of the base line and in the fourth column of townships east of the principal meridian. Township "B" is referred to as Township 2 South, Range 3 West for the same reason.[3] Every township is described with reference to the intersection of the meridian and the base lines. It is so many squares or townships north or south of the base and so many squares or ranges east or west of the principal meridian, Legal descriptions based on the Cimarron Meridian are all north of the base line and ranges are all east of the principal meridian.

Once the township and range lines are laid out, each of the squares is further divided into 36 sections (see Figure 3). Numbering always begins with one in the upper right hand corner and proceeds in a serpentine fashion to the bottom right hand corner of the township to number 36.

The Section. Once the section lines are designated, the half mile points are located so as to divide the section in four quarters. These quarters may be further divided if desired. Figure 4 shows a typical section divided into various sized tracts.

Theoretically, a section contains 640 acres and, of course, a quarter section has one fourth of 640 or 160 acres. However, because of the curvature of the earth and thus the convergence of the range lines some sections are not exactly 640 acres.[4] Sometimes surveying errors result in a discrepancy. Sometimes, you find discrepancies in size when the

3 Lines parallel to the base are referred to as township lines, those parallel to the meridian are referred to as range lines.

4 The convergence of the lines in southern Oklahoma is about 32 feet, in northern Oklahoma about 37 feet per six miles. In this State, as we go north from the base line we will find correction lines every 24 miles. This is the point where surveyors again measured the distance to the principal meridian and set out about 200 feet more or less so that the range lines were again a full six miles apart. This is why we see jogs in north and south section line roads at regular 24-mile intervals.

survey of public lands joined Indian lands. For example in Payne County, land was originally surveyed up to the southern border of the Cherokee Outlet. Then, when the "Outlet" was surveyed, the newly laid out sections along this border were odd sized and correction lots are found along the south side of the township.

The normal expectation; however, is that all sections will be full sections except those on the north and west sides of a township. Surveyors generally were instructed to confine any size deficiencies in the township to sections 1 through six and seven, 18, 19, 30 and 31. Even in these sections, shortages were to be confined to the outer borders of the section. See Figure 5. These fractional 40 acre tracts are called "lots" or "government lots".

Lots also will be found along rivers. The Oklahoma Statutes (Title 64 No. 290) declare that all streams of two chains (132 ft.) and over between the mean high water marks are State property. Private ownership stops at these lines. When such a stream cuts into any quarter-quarter to less than 40 acres it is designated as a "lot" and is assigned a number. The numbering system is not uniform in Oklahoma, but generally numbering starts farthest up the river on the right bank then down the river to the section line and back up the left bank. Figure 6. Numbering starts anew in each section. If there are "correction lots" in the section, these lots are numbered first and then the numbering is continued along the river.

Correction lots usually closely approach 40 acres in size, but a river lot may range from just a few acres to nearly 40 acres. In any event, a legal description which includes a lot, e.g. Lots 5 and 6 and S1/2 NE1/4, should warn anyone that the sale includes some fractional 40 of unknown size. Usually an instrument referring to such a tract will mention the acreage, but not always. To discover the acreage one can examine the official plat map usually found in the County Clerk's office and/ or the County Engineer's office.

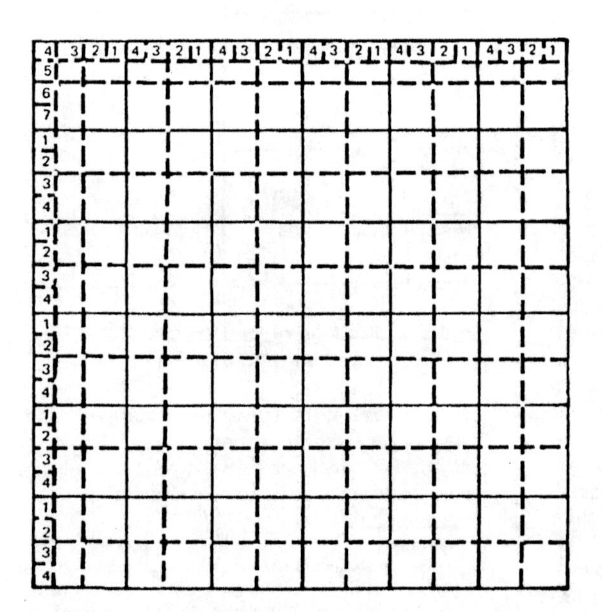

Figure 5. Sections of less than one mile square are usually on north and west sides.

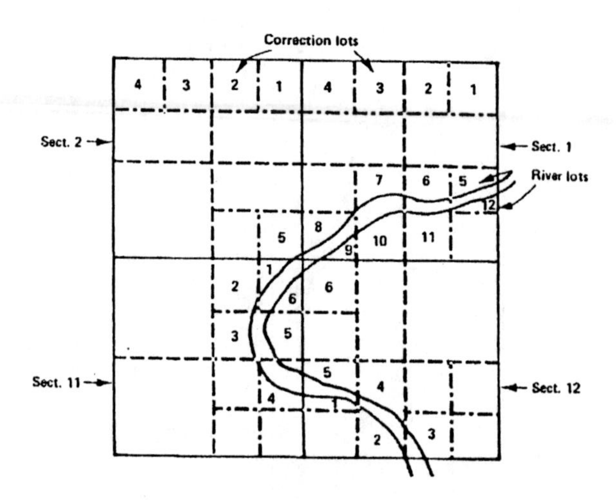

Figure 6. Lot numbers usually run downstream to section line then back upstream to line.

In writing a legal description, care must be taken that it actually describes the ownership tract. In describing a tract, the quarter section is really the key. That is, a tract usually is a quarter section, a part of that quarter, or a quarter plus other land. For example any tract smaller than a quarter section is always described as being a part of the quarter in which it is found; N1/2 SW1/4 Sect. 18 or SW1/4 SW1/4 NW1/4 Sect. 20. The final letter combination in the above descriptions designate in which of the four quarters of section the tract is found. In the first case we learn that the particular tract is the North half of the southwest quarter of Section 18. To find the land described SW1/4 SW1/4 NW1/4 Section 20 we know first that it is a tract somewhere in the NW1/4 of Section 20. But where? The legal description shows it to be somewhere in the SW quarter of the NW quarter but precisely it is the SW 10 acres of the SW 40 acres which lies in the NW 160 acres of the section.

Any time two letters appear together (NW or SW or SE) we know it indicates one fourth of something. So if we see a two letter combination followed by a section number then we know it is one fourth of the section. Thus any two letter combination is one-fourth of whatever follows.[5] If we see one letter such as N, it may or may not also be followed by 1/2 or the super-script^2. Whether the 1/2 or ^{2} follows the single letter it always means 1/2 of whatever follows.[5]

If you wish to describe a tract consisting of more than one parcel in a particular quarter, he must describe them separately. For example a 120 acre tract consisting of an 80 and a 40 acre parcel would be described in two parts even though they adjoin each other in the same quarter. Such a description might read: N1/2 SW1/4 *and* SW1/4 SW1/4 Section 10, etc. The word, or the sign meaning, "and" must separate the description of these two parcels. Of course, any tract which lies in two quarters of the section, must each indicate the quarters in which it lies, e.g. S1/2 NW1/4 *and* N1/2 SW1/4. A description such as N1/2 and S1/2 SW1/4 Section 10 *does not* describe the two halves of the SW1/4. The insertion of the word "and" means we are beginning the description of another

5 Unless the word or sign for "and" follows.

tract. When we see the word or sign for "and" we next read the section number. In the above illustration, we have described the N1/2 of Section 10 and the S1/2 of the SW1/4 of Section 10, totaling 400 acres.

Also, you can designate two quarters at one time if both are in the same half of the section; e.g. S^2 Section 8 is the same as SW1/4 *and* SE1/4 of Section 8; or E^2 Section 8 = NE1/4 *and* SE1/4 Section 8. In Figure 7, we show a farm blocked in which might be a typical unit. The farm comprises 600 acres more or less and would be described as follows: Lots 1 and 2 and S1/2 NE1/4 and SE1/4 and S1/2 SW1/4 Section 1, and S1/2 SE1/4 Section 2, and N1/2 NE1/4 Section 11, and NW1/4 NW 1/4 Section 12, all in Twp 16 N Rge. 14 W I.M.

T 16N R 14W

6	5	4	3	2	
7	8	9	10	11	12
18	17	16	15	14	13
19	20	21	22	23	24
30	29	28	27	26	25
31	32	33	34	35	36

Figure 7. Tract smaller than a quater section is described as part of quarter where it is found.

You can not only determine the location, but the size by reading the description. We know the foregoing tract is in the main body of Oklahoma because it is based on the Indian Meridian. We know further that the land is in a township which is 16 townships (between 90 and 96 miles) north of the base line and 14 ranges (between 78 and 84 miles) west of the Indian Meridian; this would place the land in the south east corner of Dewey Co. Then by the section numbers, one, two, 11, and 12, we can locate the land within the township, and the description tells us in which parts of these sections the land is found.

The size or acreage can be determined from the legal description if one remembers that a section has 640 acres. Read the description backward; for example NW1/4 NW1/4 Section 12 is one fourth of one fourth of 640 acres or *40 acres* N1/2 NE1/4 Section 11 is one-half of one fourth of 640 acres or *80 acres.* S^2SE Section 2 is one-half of one-fourth of the 640 acres in Section 2, or *80 acres.* S^2SW Section 1 is 80 acres. SE of Section 1 is one-fourth of 640 acres or *160 acres.* The S^2 NE Section 1 is *80 acres,* and lots 1 and 2 we know each to be about 40 acres in size or *80 acres more or less.* The individual acreages, when added together, equal 600 acres more or less.

For one actually to get to a legally described tract of land, he would need a general highway map of a county such as those produced by the Oklahoma Department of Highways.[6] These maps are revised each year and show every section in the county, the townships and ranges, and the type of road, if open, on every section line. Roads other than section line roads are also shown as are towns and cities and other man-made or natural features. To use such a map one merely selects a starting point and by counting section lines or reading his speedometer he drives to his destination. Such a map may be purchased from the Oklahoma Department of Highways for a nominal sum.

---

6 County maps also can sometimes be purchased from the County Engineer. Some Abstract Companies have county maps which can be used to find one's way.

Credit is extended to Loris A. Parcher and Clint E. Roush, former staff, who were authors of the original manuscript.

Issued in furtherance of Cooperative Extension work, acts of May 8 and June 30, 1914, in cooperation with the U.S. Department of Agriculture, Samuel E. Curl, Director of Cooperative Extension Service, Oklahoma State University, Stillwater, Oklahoma. This publication is printed and issued by Oklahoma State University as authorized by the Dean of the Division of Agricultural Sciences and Natural Resources and has been prepared and distributed at a cost of $276.00 for 1,500 copies. #1262 0398 ADH

16. Sketch out and label the legal description(s) in question 15 (see [illegible] for proper identification of sections).

EXERCISE 5

Soil Erosion by Water and Wind

OBJECTIVES

1. *To observe and quantify soil erosion by water*
2. *Learn what factors contribute to the soil erosion by water*
3. *Learn what factors contribute to the soil erosion by wind*
4. *Identify what factors can be used to decrease soil erosion by water*

INTRODUCTION

Erosion is a natural process that begins as soon as geologic uplift has occurred. If soil erosion is slower than soil formation, then a deep soil profile can develop. Of primary concern is when soil erosion occurs faster than soil formation. This is called accelerated erosion and is often enhanced by man's activities. Soil erosion is a major contributor to declining soil productivity and can potentially threaten the stability of a civilization. Ancient civilizations such as Mesopotamia (modern day Iraq) declined in power as the region's soil productivity decreased due to accelerated erosion enhanced by agricultural production. In United States the importance of minimizing soil erosion was brought to light during the "Dust Bowl" in the 1930's when the effects of agricultural production and several years of drought combined to produce one of the largest and most destructive periods of wind erosion in the nation's documented history.

Soil erosion can be defined as the detachment, transport, and deposition of soil materials. Water and wind are primary energy forces driving natural and accelerated erosion. For each energy force, there are three types or processes of erosion.

Water Erosion

The three types of water erosion are **sheet**, **rill**, and **gully**. Sheet and rill erosion are responsible for most of the damage caused by water erosion. Sheet erosion is the uniform loss of soil from a given area. Evidence of sheet erosion is not always obvious upon visual inspection of a field, but its damages can be substantial as whole "layers" of soil are stripped from the area. Rill erosion begins when runoff water begins to concentrate and cut small erosion channels into the soil. These small channels concentrate the runoff in a given area, increasing the water's energy and capability for erosion. An individual rill does not seem very destructive, but the total sum of erosion from all the rills in a field can be significant. The channels left behind from rill erosion can be removed with normal tillage operations, but if left unattended may develop into gullies. Gully erosion occurs in large channels that cannot be removed by normal tillage operations. Gullies have the potential to cause serious erosion problems and are the most visible sign of erosion on the landscape.

Soil loss has a significant impact on soil productivity. Therefore, it is advantageous to estimate the potential amount of soil that may be lost from a field due to water erosion. Several factors influencing water erosion including soil, climatic, landscape and management factors have been quantified in the **Universal Soil Loss Equation (USLE)** (Eq. 5.1). Using soil loss estimates for individual factors, the USLE can be used to estimate the yearly soil loss due to water erosion for a given area and management scheme.

$$A = R * K * LS * C * P \qquad \textbf{Eq. [5.1]}$$

Where:

- A = Predicted soil loss (ton/A-yr)
- R = Rainfall or climatic factor (100 ft-ton/A-yr)
- K = Soil erodibility factor (ton/100 ft-ton)
- LS = Topography factor
- C = Cropping factor
- P = Support practices

The climatic factor (R) takes into account the quantity and intensity of the rainfall at a given location based on historical data. For each county an R factor has been estimated and can be obtained from tables or maps. In Oklahoma for example, R factors range from 100 to 320 in Cimarron and McCurtin counties, respectively.

The soil erodibility factor (K) takes into account the susceptibility of a soil to erode. Generally, as a soil's silt content increases it's susceptibility to water erosion increases due to silt's small size and non-cohesive properties. Increasing sand and clay contents typically decrease a soil's susceptibility to water erosion, as sand particles are too large to move far before settling out and clay particles tend to stick together and form aggregates too large to be carried by moving water. Soil erodibility factors for soils found in each state and can be obtained in table format from the NRCS.

The topography factor (LS) is a combination of the length and steepness of the slope. Typically as a slope's steepness and length increase, so does potential water erosion due to the increasing energy of runoff water. As with the previous factors, the LS factors have been calculated and are available in table form. These first three factors (R, K, and LS) cannot be controlled by man, and so producers must work within their constraints.

Producers have much more control over the final two factors. The C factor takes into account the crop that is grown and how the crop residue is managed. Soil losses due to water erosion can be decreased by growing crops with fibrous root systems that are planted in narrow rows and by leaving crop residues at the soil surface. These cropping strategies provide above-ground vegetation and residues that strip moving water of its energy, thereby decreasing the potential for soil detachment and transport. Overall, grassland, timber, and pasture crops provide the best protection from water erosion, due to the abundance and density of vegetation and residues.

Producers also have control of the support or structural practices that they implement within a field. These P factors include row orientation, strip cropping, and terraces or other improvements to reduce erosion by absorbing the energy in runoff waters. By combining these climate, soil, topographic, and management factors, we can predict soil loss due to the erosion in an environment using the data in Tables 5.1 to 5.5.

Wind Erosion

Wind erosion processes include soil **creep**, **saltation** and **suspension**. Soil creep occurs as sand-sized soil particles are rolled and bounced short distances along the soil surface. Soil particles moved by creep typically remain in the field from which they originate. Saltation occurs when fine sand and coarse silt-sized particles are lifted, carried, and deposited from one location to the next. When saltating particles fall back to the ground, the energy from their deposition helps lift other particles into the air, thereby accelerating wind erosion. These saltating particles move only a few centimeters above the soil surface, but cause serious erosion problems. Saltating particles often collect in ditches and behind buildings or other obstructions to the wind. Suspension is the detachment, transport, and delayed deposition of clay and fine silt sized particles. These particles can stay suspended in the air for a considerable period of time and cause serious dust storms. Soil particles moved by suspension can be deposited hundreds to thousands of miles from the place of origin.

Erosion Control

Reducing the rates of accelerated erosion is an important part of soil management. By reducing soil loss, the physical, chemical and biological properties of the soil can be maintained. The keys to reducing soil erosion are to minimize the amount of water and wind energy that reaches the soil surface and minimize soil erodibility. Both objectives can be accomplished by maintaining soil cover in the form of growing vegetation or plant residue when potential erosive forces are greatest. Plant residue and vegetation can significantly decrease water erosion by reducing each raindrop's potential for soil detachment by transferring the kinetic energy in the raindrop to the residue or standing vegetation and by reducing the velocity and energy in moving water. For vind erosion, vegetative cover reduces wind velocity at the soil surface thus reducing the energy transferred to ie soil. In addition to reducing the energy in wind and water, plant residues also maintain or increase the ysical, chemical and biological properties of the soil, which help to minimize a soil's erodibility.

Table 5.1. R factors for selected Oklahoma counties

County	R Factor
Cimarron	100
Choctaw	300
Haskell	280
Osage	260
Woodward	160

Table 5.2. K factors for selected soils in Oklahoma counties

Soil	K Factor
Lincoln	0.24
Pratt	0.17
Vernon	0.37
Woodward	0.37
Yahola	0.20

Table 5.3. Topographic factor (LS) for selected combinations of slope and length

Slope	15	30	45	60	90
%			m		
2	0.16	0.20	0.23	0.25	0.28
4	0.30	0.40	0.47	0.53	0.62
6	0.48	0.67	0.82	0.95	1.17
8	0.70	0.99	1.21	1.41	1.72
10	0.97	1.37	1.68	1.94	2.37
12	1.28	1.80	2.21	2.55	3.13

Table 5.4. Crop management (C) values for different cropping rotations in Oklahoma

	Conventional tillage	
Crop Sequence	no residue	residue left
continuous wheat	0.70	0.25
continuous corn	0.85	0.65
continuous grain sorghum	0.75	0.50
continuous peanuts	0.45	0.25
grass/legume pasture	----	0.01
continuous cotton	0.48	0.25

Table 5.5. P factors for contour and strip cropping at different slopes and the terrace subfactor at different terrace intervals. To determine P value, the contour or strip cropping factor and the terrace subfactor for a terraced field must be multiplied.

			Terrace subfactor		
slope (%)	Contour P factor	Strip cropping P factor	terrace interval (m)	closed outlets	open outlets
1-2	0.60	0.30	<33	0.5	0.7
3-8	0.50	0.25	33-34	0.6	0.8
9-12	0.60	0.30	43-54	0.7	0.8
13-16	0.70	0.35	55-68	0.8	0.9
17-20	0.80	0.40	69-90	0.9	0.9
21-25	0.90	0.45	>90	1.0	1.0

Assume 10-year single storm erosion index of 130, moderate ridge height and a row grade of 0.5%.

REVIEW QUESTIONS

1. Why is it extremely difficult to alter the K factor of a soil?
2. How does growing vegetation reduce soil erosion by wind and water?
3. Which will do more to reduce soil erosion, building a terrace or maintaining crop residue on the surface?
4. How does crop residue affect a soil's physical, chemical and biological properties and reduce soil erodibility?

PROCEDURE

1. Choose a partner for this exercise. Your team will be assigned a treatment combination from Table 5.6.
2. Calculate the energy of your raindrops. To do this, fill in Table 5.7 and use Eq. 5.2:

$$E = \frac{1}{2}m\upsilon^2 = mgx \qquad \textbf{Eq. [5.2]}$$

Where:

- $E = \text{energy}\left[(J) = \frac{kgm^2}{s^2}\right]$
- $m = \text{mass of a raindrop (kg)}$
- $v = \text{velocity of a raindrop}\left(\frac{m}{s}\right)$
- $g = \text{gravitational constant}\left(\frac{9.81m}{s^2}\right)$
- $x = \text{distance (m)}$

Table 5.6. Treatment combination for soil erosion by water

Treatment	Soil Cover	Slope	Burette Height (cm)
1	bare	5 %	15
2	residue	5 %	15
3	bare	10 %	15
4	residue	10 %	15
5	bare	5 %	30
6	residue	5 %	30
7	bare	10 %	30
8	residue	10 %	30

3. Calculate the height needed to raise the back of the tray in order to achieve the assigned slope (Eq. 5.3). Record the data in (Table 5.8).

$$\text{slope} = \frac{y_2 - y_1}{x_2 - x_1} \qquad \textbf{Eq. [5.3]}$$

4. Select a soil sample from those prepared and place it in the tray behind the barriers.
5. Apply the residue treatment if required. Residue should be added on the soil surface to a uniform depth of 1-cm.
6. Place a beaker under the drain opening in the tray to collect runoff. Record runoff data in Table 5.9.
7. Set the burette so that it delivers 2 drops of water per second and let it drip for 1-hr. Be sure to refill the burette to maintain a water level within 2 cm of the top of the burette.
8. After 1-hr., stop the burette and finish data collection. Put the collection beaker in the oven until next week when a dry weight on the sample will need to be collected.
9. Compile the class data and record it in Table 5.10.

EXERCISE 5

Soil Erosion by Water and Wind
Hand-In

Name:________________________

Lab Sec.________

Points__________

Your Treatment No. ______________

Which treatment would you expect to have the most erosion? Why?

____

____

____

____

Table 5.7. Mass of a rain drop.

	mass (g)
beaker	
Beaker+20 drops	
Average mass of a drop	
Energy of a rain drop (J)	

Table 5.8. Slope calculation for soil erosion.

Your Slope = _______________	
Tray measurements	**(cm)**
$x_2 - x_1$ (rise)	
$y_2 - y_1$ (run)	

Table 5.9. Soil Runoff measurements.

Sample	**mass (g)**
Beaker	
Beaker+soil	
Soil Loss	

Table 5.10. Class soil erosion data.

Treatment	**Energy of a drop (J)**	**Soil Loss (g/hr)**
1		
2		
3		
4		
5		
6		
7		
8		

1. What five factors contribute to soil erosion by water?

2. Of the factors in Question One, which three did we evaluate in this experiment?

3. Which factor evaluated in this experiment had the most impact on soil erosion? (You will need the erosion data from all treatments to answer this question.)

4. Which has the most impact on soil erosion by water: raindrop impact or runoff? Why?

5. What properties of silt make it the most susceptible particle size to erosion?

6. As a producer what could you do to most effectively minimize soil erosion by water? By wind?

7. What are the three driving factors behind soil erosion? Which factor is the most important?

8. What equation that is used to predict soil erosion by water? Why is this equation only used to estimate erosion, not calculate true erosion?

9. What are the three types of erosion by wind?

10. Calculate the expected soil loss from a continuous sorghum crop grown in Haskell County. The soil in the field is a Vernon series. The field has a length of 45 m on a slope of 4% and is conventionally farmed although not on a contour. At the end of the cropping season the sorghum stubble is plowed in so no residue remains at the surface. (NOTE: If there is no information about a factor assume it is equal to one).

11. What is the expected soil loss in the field above if the following conditions are implemented?

    a. Residue is left on the field.

    b. Residue is left on the field and it is farmed on a contour.

    c. Residue is left on the field, contour farming is used and open outlet terraces are installed.

    d. No P factors are added and the crop is changed to a grass/legume pasture.

EXERCISE 6

Soil Textural Analysis

OBJECTIVES

1. *Gain an appreciation of the physical properties of individual soil particles (i.e. size, surface area relationships)*
2. *Learn how to use the USDA textural triangle*
3. *To determine particle size distribution (texture) of soil using the hydrometer method*
4. *Understand Stoke's law and how it describes the settling of a soil particle*
5. *To determine soil textural class of soil using the texture-by-feel method*

INTRODUCTION

Soil texture (Figure 6.1) is the ratio of sand, silt and clay sized particles in a particular soil. The terms sand, silt and clay do not refer to different minerals, but rather to different sizes of particles. Sands are particles between 2 and 0.05 mm in diameter, silts are particles between 0.05 and 0.002 mm in diameter, and clays are less than 0.002 mm in diameter. Soil texture is considered a basic property of soils and soil texture can have a significant impact on several other soil properties such as water movement, water content, pore size, aeration and aggregation. Soil texture is considered a very stable soil property due to the extremely slow rate at which soil particles weather or decrease in size. On the human time scale the only way that soil texture can be altered is through flood or erosional processes or by adding a significant quantity of a contrasting soil separate.

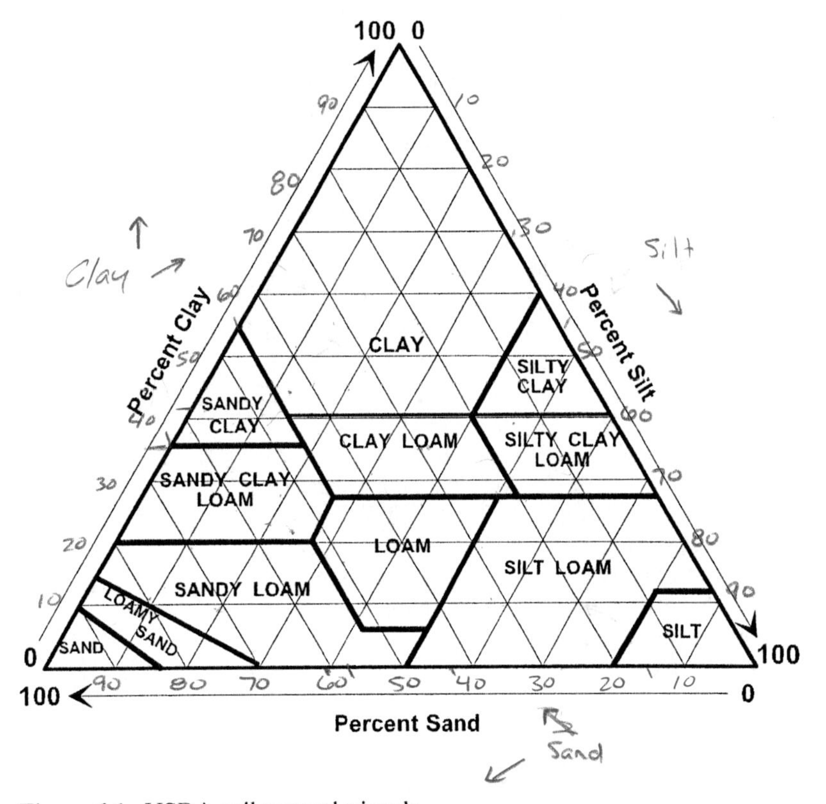

Figure 6.1. USDA soil textural triangle.

Laboratory Analysis

Determination of soil texture can be done in a laboratory setting or in the field. In the laboratory, **particle size analysis** (PSA) is conducted by suspending a soil sample in water and measuring the rate at which the soil separates settle out as described by Stoke's law (Eq. 6.1). Stoke's law states that the settling rate (v) of a particle is a result of gravity and is proportional to the square of the particle radius (r^2), the difference between the density of the particle and the density of the suspending solution (water), and inversely proportional to the viscosity (η) of the suspending solution. The equation assumes that the soil particles are individual spheres with a particle density of 2.65 Mg m^{-3}. Because the density and viscosity of water vary with temperature corrections to Stoke's law are needed when the temperature of the suspension changes during PSA.

$$\text{Stoke's Law}: v = \left[\frac{2}{9} * \left(\frac{(D_p - D_w)gr^2}{\eta}\right)\right] \qquad \textbf{Eq. [6.1]}$$

Where:

v = Velocity of fall (cm s^{-1})
D_p = Particle density (g cm^{-3})
D_w = Water density (g cm^{-3})
g = Acceleration due to gravity (cm s^{-1})
r = Radius of particle (cm)
η = Absolute viscosity of water (g $cm^{-1}s^{-1}$)

To ensure that individual soil particles are acting independently during PSA, samples are dispersed both chemically and physically. Chemical dispersion is accomplished by adding a dispersing agent to the suspension. The dispersing agent acts to repel the clay particles from each other, just as like poles of a magnet repel each other, thereby destroying soil aggregates and inhibiting aggregate reformation during the analysis. Physical dispersion is achieved by suspending the sample and dispersion solution in water and mixing at high speed for several minutes. Simply, physical dispersion breaks apart the soil aggregates and the dispersing agent helps keeps the aggregates from reforming. Another factor that must be considered is the organic matter in the sample. Soil texture is based on the mineral fraction of the soil, therefore organic matter must be removed prior to analysis. This is usually done through the oxidation of organic matter with 30% hydrogen peroxide.

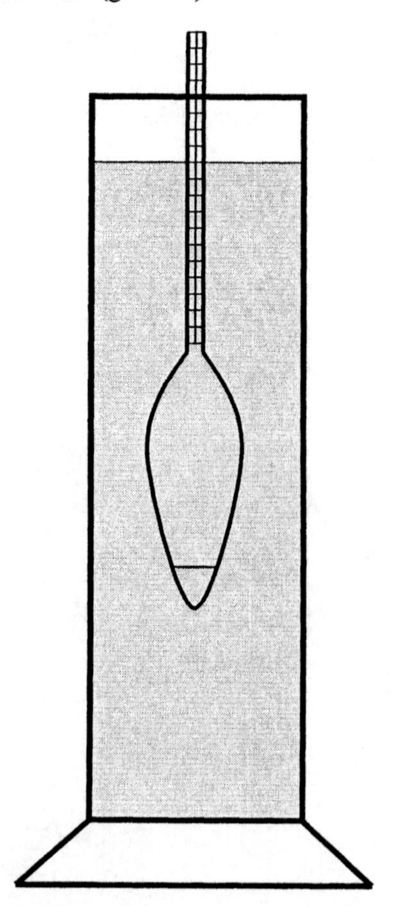

Figure 6.2. Cylinder for the determination of particle size distribution by the Bouyoucos hydrometer method.

After dispersing the sample, the soil solution is transferred to a graduated cylinder (Figure 6.2), brought to volume, and the rate of settling is measured with a Bouyoucos hydrometer. The hydrometer method is based on Stoke's law and that fact that when more particles are in suspension, the more buoyant the hydrometer and the higher the hydrometer will float in the suspension. Over time, as more particles settle out of suspension (sands fall out first, then silts, then clays), the hydrometer becomes less buoyant and therefore, floats lower in the suspension. Hydrometer readings are taken at specific times to measure the selected soil separates in g of particles L^{-1}. The hydrometers have been calibrated at 20°C; therefore the temperature of the suspension must be measured at each reading. Once PSA measurements have begun it is very important that the cylinders not be disturbed, as any disturbance will alter the rate of particle settling and result in incorrect readings.

Field Analysis

It is not always convenient (time and budget) to have particle size analysis conducted in the laboratory. Due to the characteristic properties of each soil separate, we can estimate the soil textural class by the feel of a moist soil. As you remember from exercise 1, sand particles have a gritty feel, silts feel smooth, and clays feel sticky. By using a moist sample and your fingers, you will be able to evaluate each sample for its grittiness, smoothness and stickiness. With this information the proper USDA textural class can be assigned to each sample. People who practice the texture-by-feel technique can often determine the percent sand, silt and clay to within a few percentage points.

We will determine the soil textural class for our unknown samples using both the hydrometer and texture-by-feel methods.

Review Questions

1. What will happen to the textural results of particle size analysis if soil organic matter is not eliminated?
2. Why do we use Stoke's equation for particle size distribution?
3. Why is it important to know the percent sand, silt, and clay in a soil?
4. How does soil texture affect aggregate formation?
5. Why must the soil aggregates be dispersed?

References

Day, P.R. 1965. Particle fractionation and particle-size analysis. p. 545-567. In C.A. Black et al. (ed.) Methods of Soil Analysis Part I. Agron. 9:545-567.

Jacobs, H.S. and R.M. Reed. 1964. Soils Laboratory Exercise Source Book. American Society of Agronomy. Madison, WI.

Reed, R.M. 1986. Fundamentals of Soil Science Laboratory Manual. Oklahoma State University, Stillwater, OK.

Thein, S. J. 1979. A flow diagram for teaching texture-by-feel analysis. J. Agric. Educ. 8:54-55.

PROCEDURE

A. Bouyoucos Hydrometer Method

1. Weigh out 50g (plus or minus 0.5g) of each soil in a weighing cup and record data in Table 6.1. Follow the procedure below for both soils.
2. **Quantitatively** transfer (move all the soil) the sample from the weighing into a dispersing cup (metal cup that is used on the mixer). Use deionized water from a wash bottle to rinse the weighing cup and ensure that all soil has been transferred.
3. Fill the dispersing cup to within 4 inches of the top with deionized water.
4. Add 10 mL of the dispersing agent, sodium hexametaphosphate and place the dispersing cup on the mixer. Turn the mixer on low and allow it to run for 5 minutes.
5. **Quantitatively** transfer the contents of the dispersing cup to a 1000 mL graduated cylinder and adjust to the suspension to 1000 mL using deionized water. To stir the solution, use a plunger or place the palm of your hand firmly over the top of the cylinder and invert (180 degrees) several times until no sediment adheres to the bottom.
6. Place the cylinder in an upright position and start timing **immediately**. Take your first hydrometer reading at 40 seconds. To allow the hydrometer to equilibrate, slowly place the hydrometer in the cylinder at 20 seconds. At 40 seconds record the hydrometer reading in Table 6.1 by determining where the hydrometer scale intersects with the surface of the suspension. Remove the hydrometer without disturbing the cylinder contents. Next, **hold** a thermometer in the suspension for 30 seconds. Record the temperature of the suspension. Once the hydrometer and temperature readings are completed, do not disturb the cylinder. Results will be inaccurate if the contents are restirred.
7. After 1 hour, repeat step 6. (This will be an estimate of what should be a 2 hr. reading for the silt separates).
8. Calculate the percent sand, silt, and clay using Eq. 6.1 to 6.6 and record the data in Table 6.1.

Calculations

For temperature readings greater than 20°C the corrected reading is:

$$\text{Corrected reading} = \text{hydrometer reading} + \left[0.36\frac{\text{g}}{\text{L}}(\text{measured temperature - 20})\right] \qquad \textbf{Eq. [6.1]}$$

For temperature readings less than 20°C the corrected reading is:

$$\text{Corrected reading} = \text{hydrometer reading - } \left[0.36\frac{\text{g}}{\text{L}}(20 - \text{measured temperature})\right] \qquad \textbf{Eq. [6.2]}$$

$$\%\,\text{Silt} + \text{Clay} = \left(\frac{\text{corrected 40 s reading}}{\text{mass of dry soil}}\right) * 100 \qquad \textbf{Eq. [6.3]}$$

$$\%\,\text{Clay} = \left(\frac{\text{corrected 1 hr reading}}{\text{mass of dry soil}}\right) * 100 \qquad \textbf{Eq. [6.4]}$$

$$\%\,\text{silt} = (\%\,\text{silt} + \text{clay}) \text{ - } \%\,\text{clay} \qquad \textbf{Eq. [6.5]}$$

42.1 - 41.4

$$\% \text{ sand} = 100 - (\% \text{ silt} + \text{clay}) \qquad \textbf{Eq. [6.6]}$$

B. Texture-by-Feel Method

1. To determine soil textural class with the texture-by-feel method, first practice with known samples supplied by the instructor to "calibrate" your fingers.
2. Follow the procedure outlined in Figure 6.3 for two different known samples.
3. After you have practiced with this technique, now give it a try with the unknown samples.
4. Put your estimate of the textural class in Table 6.1.

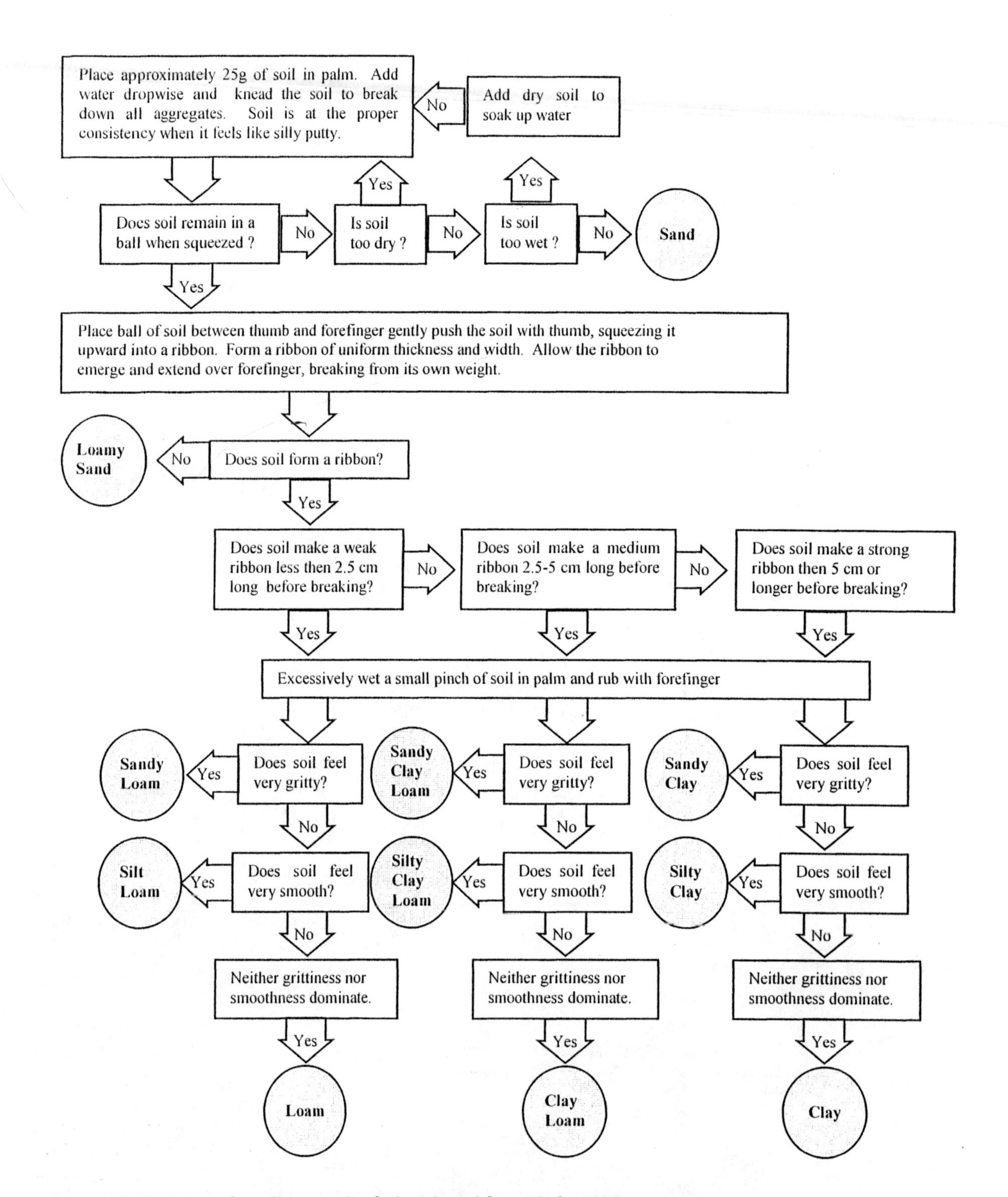

Figure 6.3. Flow chart for soil texture-by-feel. Adapted from Thein, 1979.

EXERCISE 7

Soil Physical Properties of Density and Pore Space

OBJECTIVES

1. *To determine the bulk density of soils*
2. *To determine the particle density of soils*
3. *To recognize the relationship between bulk density, particle density, and pore volume of a soil*
4. *Be able to calculate the pore volume of a soil*
5. *Be able to calculate the air volume and water volume of a soil*

INTRODUCTION

Soil physical properties of **particle density, bulk density** and **pore space (volume)** have significant impacts on many other physical and chemical soil properties. Particle density is considered a constant property in soils and is a function of the minerals present in a soil. Bulk density is the density of the soil matrix (solids and voids) and directly impacts the total volume of pores in the soil. In turn, the volume and size of pores within the soil matrix affects many aspects of soil productivity including water movement and soil aeration.

PARTICLE DENSITY

Particle density (**D_P**) is a measure of the weight of a dry soil (**W_S**) per unit volume of solids (**V_S**) in the soil (Eq. 7.1). Particle density is a function of the minerals that comprise a particular soil and can range from 2.0 to 7.6 Mg m^{-3} (Table 7.1). However, due to the predominance of quartz and other aluminosilicate minerals in soils (Table 2.1), the range for particle density is typically from 2.6 to 2.7 Mg m^{-3}. If the particle density of a soil is not measured it is often **assumed** to be 2.65 Mg m^{-3}.

Table 7.1. Particle density (D_P) of selected minerals found in soils of Oklahoma

Mineral	D_P (Mg m^{-3})
bauxite (aluminum ore)	2.00
quartz and aluminosilicates	2.60
calcite	2.72
hematite (iron ore)	5.30
galena (lead ore)	7.60

$$D_P = \left(\frac{\text{mass of dry soil}}{\text{volume of solids}} = \frac{W_S}{V_S} \left(\frac{\text{Mg}}{\text{m}^3} \text{ or } \frac{\text{g}}{\text{cm}^3} \right) \right) \qquad \textbf{Eq. [7.1]}$$

BULK DENSITY

Bulk density (**D_B**) is a measure of the mass of dry soil to the total volume (**V_T**) of soil (Eq. 7.2). The difference between particle density and bulk density is that bulk density includes the volume of the pores (**V_P**) in the soil. A "typical" soil is composed of 50% solids and 50% pores and has an average bulk density of 1.33 Mg m^{-3}. Bulk density is a function of the ped development (aggregation) and soil texture, therefore the bulk density will change with any practice or process that alters soil aggregation, porosity soil organic matter, or the ratio of soil separates. Any action that decreases the volume of the pores in a soil (e.g. destruction of soil structure, loss of organic matter) will increase the bulk density and reduce overall soil productivity. Reduced pore volume means there is less air in the soil for root respiration, fewer voids to store water, and more stress on the plant roots to expand (grow) because of the physical pressure needed to "push" the soil out of the way. Therefore, agricultural producers, horticulturalists, foresters, or anyone interested in plant growth, should be concerned with minimizing soil compaction in order to maintain a suitable soil bulk density (volume of pores) for optimal plant production. In contrast, contractors or builders who want a good base for construction purposes will want to increase soil bulk density, thereby decreasing the porosity of the soil in order to minimize the shifting and settling of construction foundations. The relationship between bulk density and soil volume is used in determining soil porosity.

$$D_B = \left[\frac{\text{mass of dry soil}}{\text{total soil volume (solids + pores)}} = \frac{W_S}{V_T}\left(\text{Units} \frac{M_g}{m^3} \text{ or } \frac{g}{cm^3}\right)\right] \qquad \textbf{Eq. [7.2]}$$

PORE SPACE (VOLUME)

The total volume of a soil (**V_T**) is composed of the volume occupied by solids (**V_S**) and pores (**V_P**). The pore volume is comprised of the volume of pores that contain air (**V_A**) or water (**V_W**) as shown in Eq. 7.3. Porosity is the pore space (volumem^3) in a soil.

$$V_T = V_S + V_P = V_S + (V_A + V_W) \qquad \textbf{Eq. [7.3]}$$

We are often interested in the percent total pore space (**PS_P**) of a soil. Percent pore space is relative to the ratio of the soil pores to the total soil volume. When considered on a percentage basis, the percent total space of the soil (**PS_T**) must equal 100%. Therefore the sum of the percent solid space (**PS_S**) and percent pore space (**PS_P**) must total 100% (Eq. 7.4).

$$PS_T = 100 = PS_S + PS_P = PS_S + (PS_A + PS_W) \qquad \textbf{Eq. [7.4]}$$

Measuring the volume of the pores V_P is very difficult. Instead the volume of the solids is determined then the volume of the pores is calculated. To go from measuring V_S to calculating V_P, the relationship of D_P (does not consider V_P), to D_B (does consider V_P) is used (Eqs. 7.1 and 7.2).

$$D_P = \frac{W_S}{V_S} \text{ and } D_B = \frac{W_S}{V_S + V_P} = \frac{W_S}{V_T} \qquad \text{Eqs. [7.1 and 7.2 rewritten]}$$

When these equations are solved for the weight of solids (W_S), the following relationships are obtained.

$$W_S = D_P V_S \qquad W_S = D_B V_T \qquad \text{Eqs. [7.1 and 7.2 rewritten]}$$

Now, setting the Eq. 7.1 and 7.2 equal to each other, the ratio of V_S to V_T is determinedthus PS_S.

$$D_P (V_S) = D_B (V_T) \text{ or } \frac{V_S}{V_T} = \frac{D_B}{D_P} \qquad \textbf{Eq. [7.5]}$$

The results from Eq. 7.5 are multiplied by 100 to obtain **PS_S** (Eq. 7.6).

$$PS_S = \frac{V_S}{V_T} * 100 = \frac{D_B}{D_P} * 100 \qquad \textbf{Eq. [7.6]}$$

From Eq. 7.4, the PS_T is 100%, therefore substitution of the Eq. 7.6 into Eq. 7.4 can be solved for PS_P (Eq. 7.7).

$$PS_P = 100 - PS_S$$

or **Eq. [7.7]**

$$PS_P = 100 - \left(\frac{D_B}{D_P}\right) * 100$$

Percent pore space provides a picture of the amount of the soil volume occupied by pores. However, by knowing the percent pore space of a soil we know nothing of its pore size distribution or if the pores are filled with water or air. By determining the volume of water in a soil the volume of air can be obtained. When working with V_P and PS_P remember that units of V_P are a unit volume (e.g. cm^3 or mL) and PS_P is a percentage, therefore unitless.

REVIEW QUESTIONS

1. How do particle density and bulk density differ?
2. Why is bulk density an important property in determining soil productivity?
3. How can the bulk density of a soil be increased or decreased?
4. Why does the particle density of a soil remain relatively constant?
5. Why does the porosity of a soil depend on the bulk density and the particle density?
6. How does an idealized soil have a D_B of 1.33 Mg m^{-3}, if the rock it was formed from had a D_P of 2.65 Mg m^{-3}? What must occur?
7. Define these terms, V_T, V_S, V_P, D_P, and PS_P.
8. Why is D_P a fixed value for most soils?

REFERENCES

Beyrouty, C.A., N. Slayton and M.G. Hanson. 1992. Lab Manual for Soils. University of Arkansas.

Reed, R.M., 1986. Fundamentals of Soil Science Laboratory Manual. Oklahoma State University, Stillwater, OK.

Weil, R.R., 1981. A Laboratory Manual for General Soils. Avery Publishing Group Inc. Wayne, N.J.

PROCEDURE

A. Particle size affect on D_B, D_P, and PS_P

1. Using the idealized soil particles (beads) in the graduated cylinders, calculate the D_B, D_P, and PS_P for beads in each cylinder.
2. Use the information about each color bead in Table 7.2 to make your determinations.

B. Bulk Density (D_B)

1. Select a soil from those available. Each soil has two treatments, one is ground and the other is unground.
2. Weigh a graduated cylinder for each sample and record the weight in Table 7.3.
3. For each sample (ground and unground), fill a 100 mL graduated cylinder approximately ½ full. Do not tap the cylinder because this will compact the soil.
4. Record the volume of the **uncompacted** soil in Table 7.3.
5. Now tap the cylinder firmly on the bench until no change in volume is achieved with additional taps.
6. Record the volume of the compacted soil in Table 7.3.
7. After recording the compacted volume, place the soil in a beaker and **save** for use in determining particle density.
8. Using the weight and volume of each sample, calculate the D_B for the uncompacted and compacted samples.

C. Particle Density (D_P)

1. Fill the emptied graduated cylinders to approximately 50 mL with distilled water. Record the volume in Table 7.4.
2. To the cylinder that contains 50 mL of water, slowly add the soil saved from the determining D_B.
3. When the soil is added, make sure it does not stick to the side of the cylinder. Tap the cylinder to ensure that all of the soil is submerged in the water.
4. Let this stand for five (5) minutes to allow any entrapped air to escape.
5. After 5 minutes, record the cylinder volume in Table 7.4.
6. Calculate the D_P from the measurements of soil mass and volume of the cylinder and record the results in Table 7.4.

D. Percent Pore Space (Volume) (PS_P)

From your determinations of D_B and D_P, calculate the PS_P and record the results in Table 7.5.

EXERCISE 8

Properties of Soil Water

Objectives

1. *To understand how water in soil moves by saturated flow and measure the saturated hydraulic conductivity of soils with different textures*
2. *To understand how water in soil moves by unsaturated flow and measure that movement by capillary rise with soils of different textures*
3. *To observe how the water content of a soil can be measured in the field using tensiometers*
4. *To determine plant available water from known soil properties*

INTRODUCTION

Properties of Water

Water is needed for almost all the functions of life. Therefore, the properties of soil water are important to each of us and to the world we live in. Many of the unique properties of water result from the arrangement of hydrogen atoms within the molecule (Figure 8.1).

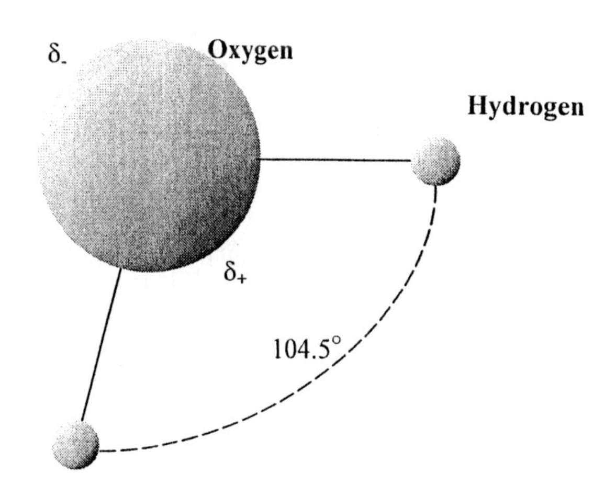

Figure 8.1. Structure of a water molecule

Water is an electroneutral molecule with positive and negative charge concentrated at opposite ends of the molecule (δ_+ and δ_-) causing it to have a polar nature. The orientation of the partial charge is an important feature in how water reacts with other water molecules and with everything around it. The polar nature of water allows for **H-bonding** or the absorption of water molecules to each other and to soil surfaces (Figure 8.2). The attractive force of water molecules to each other is called **cohesion** and attraction to other surfaces is **adhesion** or adsorption. The forces of adhesion and cohesion combined with the force of **gravity** have a significant effect on how water moves through soils.

Water Movement in Soils

There are three types of water flow in soils: **saturated, unsaturated**, and **vapor**. Saturated and unsaturated flows have the most significant effect on water flow in the soil. Saturated flow occurs when all the soil pores are filled with water and the force of gravity moves water from the soil surface downward into the soil profile. Saturated flow occurs in most soils during periods of frequent intense precipitation and although it is the primary form of water movement in wetland conditions. Saturated flow has practical applications for those who use septic systems for domestic sewage disposal. When installing a septic system, water infiltration tests called perk tests are often conducted on a soil to

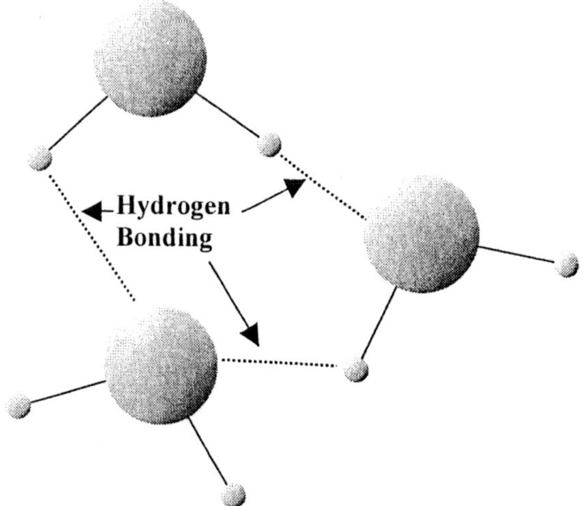

Figure 8.2. Diagram of hydrogen bonding between water molecules

make sure that the hydraulic conductivity will meet the needs of household. **Saturated hydraulic conductivity** is described in Eq. 8.1 and is demonstrated in Figure 8.3.

$$K_{sat} = \frac{V d_s}{d_w A t}$$ **Eq. [8.1]**

where:

K_{sat} ≡ Saturated hydraulic conductivity ($cm\ s^{-1}$)
V ≡ Volume of water collected during the time period (cm^3)
d_s ≡ Depth of the soil (cm)
d_w ≡ Depth of the water (cm)
A ≡ Surface area of the soil (cm^2)
t ≡ Collection time period (s)

The USDA-NRCS (Natural Resources Conservation Service) has classified the saturated hydraulic conductivities of various soil textures (Table 8.1). This classification system is useful for landowners and managers who make decisions about land use, as the hydraulic conductivity of a soil provides the users with an estimate of the drainage and flooding potential of a site. For example, a person would not want to purchase land for a waterfowl wetland that is classified as having a high saturated hydraulic conductivity, as the soil will not hold water.

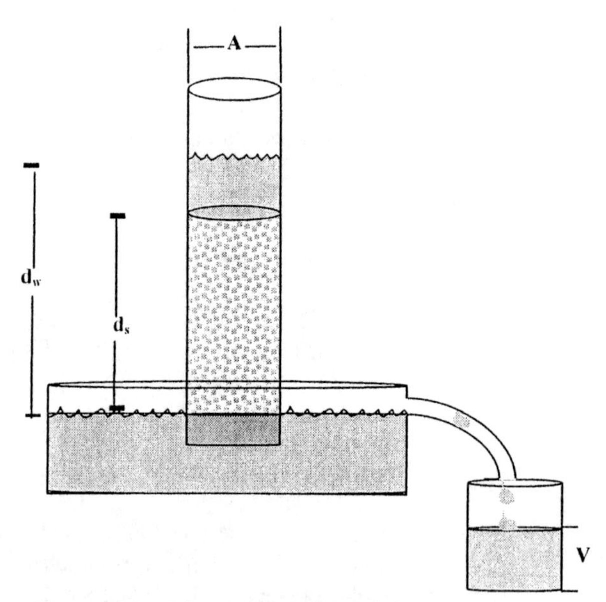

Figure 8.3. Diagram for determining saturated hydraulic conductivity of a soil

Table 8.1. USDA-NRCS classification of hydraulic conductivity for soils

Class	Rate	Soil Texture
very low	0.25 cm hr^{-1} (0.1 in hr^{-1})	clays
low	0.25 to 1.25 cm hr^{-1}	clays
medium	1.25 to 2.5 cm hr^{-1}	loams/silts
high	> 2.5 cm hr^{-1}	sands, well aggregated, silt loams

Unsaturated flow is the primary mode of water movement in most soils. The forces of **adhesion** and **cohesion** dominate the movement of water in unsaturated conditions and unsaturated flow is often referred to as **capillary movement**. Unsaturated conditions by definition, mean that some pores contain air. The macropores generally contain air, although this will vary with the moisture content of the soil. Therefore most water movement in unsaturated conditions is through the micropores due to capillary action. The rise of water in a column occurs by capillary action as described in Eq. 8.2

$$h = \frac{2T}{rgD_w}$$ **Eq. [8.2]**

where:

h ≡ Height of rise (cm)
T≡ Surface tension of water (i.e. cohesion of water at the water/air interface (71.97 dynes cm^{-1}))
r ≡ Radius (cm) of the capillary pore **(not of the soil particle)**

g ≡ Gravitational acceleration (980 cm s^{-2})
D_w ≡ Density of water (1.0 g cm^{-3})

and for most soil conditions can be reduced to:

$$h = \frac{0.15}{r} \qquad \textbf{Eq. [8.3]}$$

It is impossible to measure the radius of every pore in the soil, but by measuring the height of rise of water in a soil column we can determine the effective diameter of the largest soil pores that conduct water during unsaturated flow. The same properties that cause water to rise in a soil column, work to transport water to the plant roots.

Soil Water Content

The **soil water content** is important to soil productivity, as plants depend on the soil for their source of water. Therefore, it is important for us to quantify the amount of water in the soil and more importantly know how much is available for plant growth. The amount of water in the soil can be measured directly or indirectly. The direct method or **gravimetric method,** is a accurate way to measure the amount of water in a soil, but is time consuming. It requires that a soil sample be collected, weighed wet, then allowed to dry (generally at 105°C) for 24-hrs, and then weighed dry. Water content is calculated using Eq. 8.4,

$$\theta_W = \frac{wet_{soil} - dry_{soil}}{dry_{soil}} \qquad \textbf{Eq. [8.4]}$$

Where:

θ_W ≡ gravimetric water content
wet_{soil} ≡ wet weight of soil
dry_{soil} ≡ dry weight of soil

Percent water by weight is calculated in Eq. 8.5

$$\%H_2O_W = \theta_W * 100 \qquad \textbf{Eq. [8.5]}$$

In order to obtain a relative picture of how much water is present in a soil profile as compared to soil particles and pore space, the water content must be converted from a weight basis (θ_W) to a volume basis. This is done in Eq. 8.6

$$\theta_V = \theta_W * \frac{D_B}{D_W} \qquad \textbf{Eq. [8.6]}$$

Where

θ_V ≡ volumetric water content (cm^3 cm^{-3})
D_W ≡ density of water (Mg m^{-3})

For typical soil conditions D_W can be approximated at 1.0 Mg m^{-3}.

Volumetric water content can be converted to a percent basis using Eq. 8.7 .

$$\%H_2O_V = \theta_V * 100 \qquad \textbf{Eq. [8.7]}$$

Indirect methods of measuring soil water content include the use of the **neutron probe, gypsum blocks,** and **tensiometers.** The purpose of these measurement methods is to determine the moisture content of the soil in the field, so that management decisions can be made at that time. The neutron probe measures the rate at which a neutron travels from a source through the soil to a detector. While traveling from the source to the detector, the neutron will react with the H atoms in the soil. The more H atoms in the soil, the more water, and the longer it takes for the neutrons to reach the detector. This method must be calibrated in the lab to determine the return rate of neutrons at various moisture contents. Due to the calibration and the fact that most people do not want to work with a neutron source, this method is not widely used.

Gypsum blocks are used to measure electrical resistance as a current passes through the gypsum block (Figure 8.4). When an electrical current is passed through the block, there is less electrical resistance within the block as the soil water content increases. As with the neutron probe, gypsum blocks must be calibrated prior to use in order to establish the relationship between soil water content and resistance. For proper use, gypsum blocks should be installed at varying depths in the soil profile.

Tensiometers measure the soil's matrix potential or the attraction of water to the soil surface. Soil water content and matrix potential are directly related. As soil water content increases matrix potential increases, and the tension at which water is held to the soil surface decreases. Tensiometers must be installed in the soil so that a good contact is established between the soil and the ceramic cup. As the soil dries, water is pulled from the porous ceramic cup out of the tensiometer and into the soil. This creates a vacuum that is measured by the pressure gauge at the top of the tensiometer. Tensiometers are often installed at different depths so that the soil moisture potential can be determined at throughout the growing profile (Figure 8.5). Tensiometers are used in irrigated agriculture to determine when irrigation is needed.

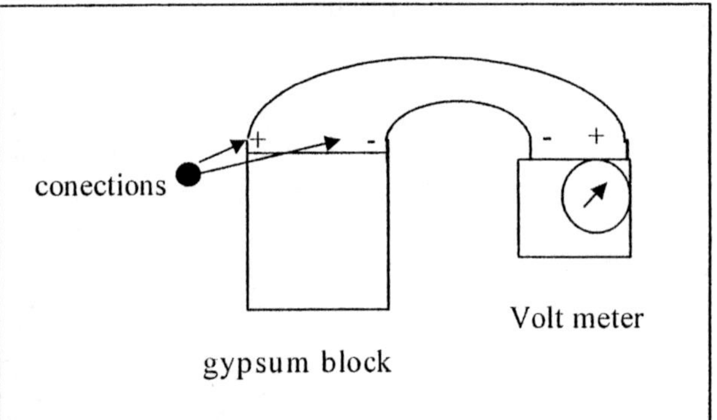

Figure 8.4. Diagram of a gypsum block used to measure the resistance of electrical current through the soil

Plant Available Water

Both direct and indirect methods of measuring water content measure the total amount of water in the soil, but not all the water in a soil is **plant available water (PAW)**. Plant available water is that water held at moisture potentials between **field capacity** (-33 kPa) and **permanent wilting point** (approximately -1500 kPa). At moisture potentials between 0 to –33 kPa, the macropores are drained quickly by gravity and roots cannot adsorb the percolating water. At moisture potentials less than –1500 kPa, the water adheres to the soil so tightly that it cannot be removed by the plant. Therefore, by knowing the amount of water held between –33 to –1500 kPa, and the depth of a soil profile the PAW of a specific soil profile can be determined (Eq. 8.8).

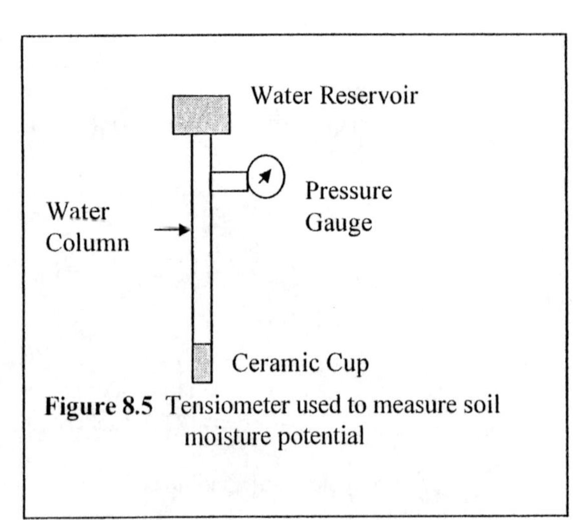

Figure 8.5 Tensiometer used to measure soil moisture potential

$$d_w = \Delta\theta_V * d_s \qquad \textbf{Eq. [8.8]}$$

Where:

$d_w \equiv$ Depth of water,
$\Delta\theta_V \equiv$ Change in volumetric water content,
$d_s \equiv$ Depth of the soil profile.

The amount of PAW stored in a profile can be expressed as acre-feet of water or acre inches of water just as in water reservoirs or impoundments

Example

The percent water by volume in a soil has previously been determined to be 35.4% at –33 kPa and it is 16.3% at –1500 kPa. How much PAW is in a soil profile that is 45-cm deep if the soil was just irrigated?

$$d_w = \frac{(35.4 - 16.3\%)}{100} * 45 \text{ cm soil} = 8.6 \text{ cm of water in the soil to a depth of 45 cm}$$

REVIEW QUESTIONS

1. How does water move in the soil and which is the primary mode of water movement in most soils?
2. Why is it important to know how much plant available water is in a soil profile?
3. What characteristic of a water molecule is the driving force behind capillary movement?
4. Why is water important for plant and microbial life?

PROCEDURE

A. Saturated Flow

1. Use the soil columns set-up by the laboratory instructor to determine saturated flow of different soil textures.
2. The columns will be set-up so that a constant level (pressure head) of water is added to the top of the soil surface.
3. The volume of water collected during the lab period will be measured.
4. With the volume of water collected during the sampling time, calculate the K_{sat} for each soil using Eq. 8.1.
5. Enter the results in Table 8.2.

B. Video of Water Movement in Soils

1. A video on water movement in soils will be shown. Many properties of water movement will be demonstrated.
2. Be prepared to answer questions on the video.

C. Unsaturated flow

1. Use the columns set up on the bench to measure the unsaturated flow in the soil.
2. Use a ruler to measure the height of rise of the wetting front in each water column.
3. Record the height in Table 8.3.

D. Soil Water Content

1. Three containers have been set up to measure the water content of a soil. The water contents of the soil by weight are approximately
   a. field capacity
   b. 20%
   c. 10%
2. Record the soil moisture potential in Table 8.4.
3. Sketch a diagram of a tensiometer.

EXERCISE 9

Soil Temperature and Atmosphere

OBJECTIVES

1. ***To observe how soil temperature is affected by soil moisture, soil texture, and organic residue***
2. ***To calculate the total energy adsorbed by soils in a given period of time***
3. ***To observe how flooding changes the soil atmosphere and the redox potential of a soil***

INTRODUCTION

Soil Temperature

Soil temperature has significant impact on many soil properties, such as chemical reactions, microbial growth, and plant growth. Chemical reaction rates increase as soil temperature increases. The Q_{10} is a rule-of-thumb that says that the rate of a chemical reaction will double for every 10°C increase in temperature. Therefore, soil temperature is important in determining the chemical dynamics of a soil and ultimately, the nutrient supply for plants. Soil temperature also affects which members of the microbial population are active at any given time and the rate of their biochemical reactions. In addition, soil temperature affects the rate of plant root growth and biochemical activity. In the spring, many seeds do not germinate until the soil warms to a certain temperature, and if the seeds do germinate their root systems will tend to grow more slowly in cool soils.

The primary source of heat energy for soils is from solar radiation. Pathways of energy adsorption within the soil are direct and indirect, with most soil heat energy coming from the direct adsorption of solar radiation and diffuse scatter radiation.

Once solar radiation is adsorbed by the soil, the increase in soil temperature is a function of several factors including soil water content, the amount of organic residue on the surface, soil texture and slope aspect. Of these factors, soil water content has the most affect on the rate at which soils warm and cool. The large specific heat of water requires large quantities of energy must be acquired in order to increase the temperature of the water. Therefore, wetter soils are buffered against temperature changes, they will warm up and cool down slower than drier soils. **Specific heat** is defined as the amount of energy (cal) needed to raise 1 g of material 1°C (Eq. 9.1). Soil solids and water have different specific heats. The specific heat for soil minerals and water are 0.2 cal g^{-1} $°C^{-1}$ and 1.0 cal g^{-1} $°C^{-1}$, respectively.

$$\text{Specific Heat} = \frac{\text{cal}}{\text{g}\ {}^{o}\text{C}} \qquad \textbf{Eq. [9.1]}$$

Therefore, the energy needed to raise 24 g of soil minerals and 5.4 g of water 1°C is given in the following example.

$$\text{Energy} = (\text{mass}_{\text{soil}} * \text{Specific Heat}_{\text{minerals}}) + (\text{mass}_{\text{water}} * \text{Specific Heat}_{\text{water}})$$

$$\text{cal} = \left(24\ \text{g}_{\text{soil}} * \frac{0.2\ \text{cal}}{\text{g}_{\text{soil}}\ {}^{o}\text{C}}\right) + \left(5.4\ \text{g}_{\text{water}} * \frac{1.0\ \text{cal}}{\text{g}_{\text{water}} * {}^{o}\text{C}}\right) = 10.2\ \text{cal}$$

There are many pathways for heat loss from soils, including reradiation, evaporation transpiration, and reflection.

Soil Atmosphere

The amount of air in the soil has a major effect on biological and chemical reactions in the soil. Soil air contains the same constituents present the atmosphere, but in different concentrations (Table 9.1). In general, soils have lower O_2 and higher CO_2 concentrations than the atmosphere due in part to respiration of plant roots and soil microbes. Respiration can be summarized by the reaction:

$$C_6H_{12}O_6 + 6O_2 \leftrightarrow 6CO_2 + 6H_2O + \text{energy}$$

Table 9.1. Average amount of compounds found in the soil and in the atmosphere

	Air	Soil
Compound	(%)	
N_2	78	78
O_2	21	20
CO_2	0.035	0.5
H_2O	20-90	95-99

Respiration is the conversion of reduced C to oxidized C resulting in the release of energy. During respiration, O_2 is used as the terminal electron acceptor thereby allowing organisms to carry out aerobic respiration. When O_2 is not present, some organisms are capable of using other compounds as an electron acceptor for alternate respiration pathways. To determine if a molecule is being used as an electron acceptor, one can measure the redox potential (Eh) of a soil in voltage (V). Compounds that undergo redox reactions do so at characteristic Eh values. Therefore, by knowing the Eh of a soil we can establish which form (i.e. oxidized or reduced) of an element is present (Table 9.2). If the Eh measurement is more positive than the characteristic Eh value, then the oxidized form of that compound or element is present. If the Eh measurement is more negative than the characteristic Eh value, then the reduced form is present. For example, a soil with a measured Eh of 0.220V would indicate that most of the manganese would be in the Mn^{2+} state but the iron would be in the Fe^{3+} oxidation state.

Table 9.2. Characteristic redox potentials (Eh) for selected compounds which are often present in the soil

Element	Oxidized Form	Reduced Form	Eh (V)
O	O_2	H_2O	0.400
N	NO_3	NH_4	0.300
Mn	Mn^{4+}	Mn^{2+}	0.250
Fe	Fe^{3+}	Fe^{2+}	0.200
S	SO_4^{2-}	H_2S	-0.500
C	CO_2	CH_4	-0.100

Review Questions

1. Why does it take longer to heat a wet soil versus a dry soil?
2. Why is soil temperature important for plant growth?
3. What are the ways heat energy is lost from the soil? Which one is most important?
4. Why does the composition of soil air differ from the atmosphere?

REFERENCES

Jacobs, H.S. and R.M. Reed. 1964. Soils Laboratory Exercise Source Book. American Society of Agronomy. Madison, WI.

PROCEDURE

A. Soil Temperature

1. Choose a partner and your group will be assigned a treatment combination from Table 9.3.
2. Add 1,000 g of soil to a tray.
3. Add the assigned water treatment (remember this is on a weight basis). Uniformly add the water to the soil and mix.
4. Place the bulb of a thermometer under the soil to a depth of 1.3 cm (0.5 in).
5. Add the appropriate residue treatment to a depth of 1.25 cm.
6. Make sure that the energy source (lamp) is positioned 20 cm above the soil surface.
7. Turn the lamp on and begin reading the thermometer in 5min. intervals for 60 min. Record the data in Table 9.5.
8. Plot the data on the graph included.
9. When you have finished collecting the data, record the total temperature change on the board.
10. Record the class data in Table 9.6.

Table 9.3. Treatment combinations for measuring soil temperature changes

Treatment	Soil	% H_2O_W	Residue
1	Organic Soil	10	Bare
2	Organic Soil	10	Covered
3	Organic Soil	20	Bare
4	Organic Soil	20	Covered
5	Sand	10	Bare
6	Sand	10	Covered
7	Sand	20	Bare
8	Sand	20	Covered

B. Soil Atmosphere

1. A demonstration is set-up to show differences in soil atmosphere.
2. Treatments for the demonstration are listed in Table 9.4.
3. The lab instructor will measure the redox potential of the soil and to be recorded in Table 9.7.

Table 9.4. Treatments for soil atmosphere demonstration

Soil	% H_2O_W	Organic Matter (% by weight)
Sand	20	None
Sand	Flooded	None
Sand	Flooded	1
Organic soil	20	None
Organic soil	Flooded	None
Clay loam	20	None
Clay loam	Flooded	None
Clay loam	Flooded	1

EXERCISE 9

SOIL TEMPERATURE AND SOIL ATMOSPHERE
Hand-In

Name:____________________

Lab Sec.________

Points________

Soil Temperature

Table 9.5. Changes in soil temperature as affected by residue and water differences

Time	Temperature (°C)	Time	Temperature (°C)
0		35	
5		40	
10		45	
15		50	
20		55	
25		60	
30		Temp. Change	

Temperature Graph

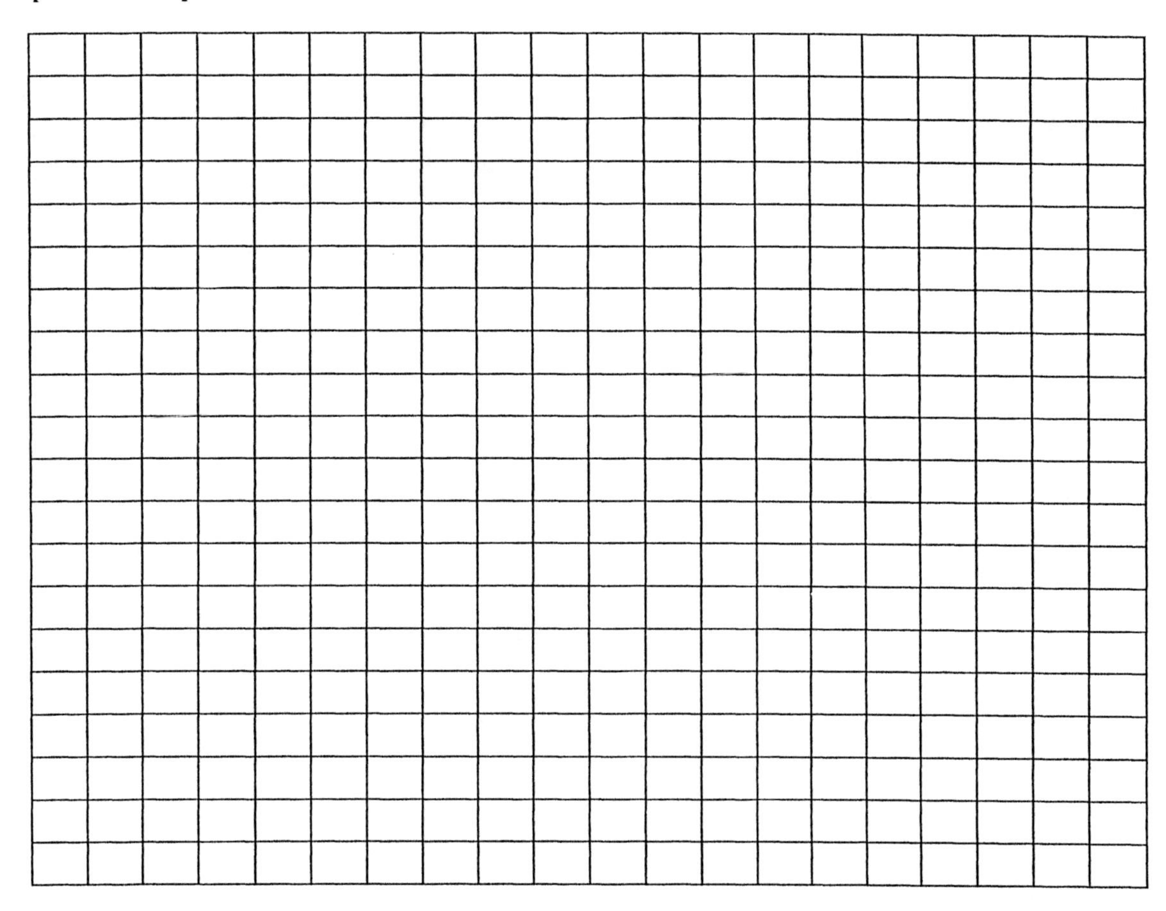

Table 9.6. Treatment combinations for measuring soil temperature changes

Soil	Water Content (%)	Residue	Temperature change (°C)
Organic Soil	10	Bare	
Organic Soil	10	Covered	
Organic Soil	20	Bare	
Organic Soil	20	Covered	
Sand	10	Bare	
Sand	10	Covered	
Sand	20	Bare	
Sand	20	Covered	

Soil Atmosphere

Table 9.7. Treatments for soil atmosphere demonstration

Soil	Water content (% H_2O_W)	Organic Matter (% by weight)	Redox Potential (mV)
Sand	20	none	
Sand	flooded	none	
Sand	flooded	1	
Organic soil	20	none	
Organic soil	flooded	none	
clay loam	20	none	
clay loam	flooded	none	
clay loam	flooded	1	

Questions and Calculations

1. Calculate the total energy (cal) that was adsorbed by your soil.

2. Which treatment had the smallest change in soil temperature?

   ____

3. Which treatment had the greatest change in soil temperature?

   ____

4. What effect did the residue have on the soil temperature change? Why?

   ____
   ____
   ____

5. What effect did water content have on the soil temperature change? Why?

   ____
   ____
   ____
   ____

6. How did flooding the soil affect the redox potential of the soils? Why?

   ____
   ____
   ____
   ____

7. How did the organic matter affect the redox potential of the flooded soils? Why?

   ____
   ____
   ____
   ____

8. Why can we assume that the average heat capacity for soil solids is 0.2cal g^{-1}_{soil}?

   ____
   ____
   ____

9. What form of the element would be expected in the clay loam soil that was moist and flooded with organic matter? Put your responses in Table 9.8.

Table 9.8. Forms of selected elements expected to be found in moist and flooded soils

	Compound	
element	**moist soil**	**flooded soil with organic matter**
O		
N		
Fe		
Mn		

EXERCISE 10

Exchange Properties of Soils

OBJECTIVES

1. *Understand the importance of exchange reactions in soils*
2. *Learn the exchange characteristics of different soil colloids*
3. *Learn the importance of cation exchange capacity*
4. *Estimate the cation exchange capacity of soils*
5. *Measure the cation exchange capacity of soils*

INTRODUCTION

Exchange Properties

Exchange properties are one of the most important attributes of soils. They are a result of physical and chemical reactions that occur at the solid-solution interface when ions in the soil solution become electrostatically bonded to charged sites on soil colloids. For plant production, exchange reactions are probably the most important reaction in soil. Without the capacity of soils to hold plant nutrients on colloid surfaces and release them into soil solution over time, our ability to grow plants for food, fiber, fuel, and aesthetics would be greatly diminished. Therefore, as the number of exchange sites increases, the productivity of the soil generally increases. There are many types of soil colloids that contribute cation exchange, each with their own set of unique properties.

Soil Colloids

Exchange properties are dependent on the types of soil colloids present. Some colloids have charges at or very near the surface that will attract and hold **ions** of the opposite charge. For most soils, the colloids contain predominately negative charges, thus attracting positively charged ions (**cations**) from the soil solution to balance their negative charge.

Soil colloids differ greatly in composition. Many are primary or secondary minerals (Exercise 2), some are amorphous compounds that have little structure, and others are formed from the decomposition of organic matter (humus). Based on the formation environment, each colloids has its own unique chemical and physical characteristics that cause it to have specific exchange properties (Table 10.1). The geographic distribution of soil colloids is a function of climate, profile development and parent material. In general, rapid and intense weathering produces kaolinite and iron or aluminum oxides respectively, whereas less intense weathering promotes formation of montmorillonite.

Table 10.1. Exchange characteristics, range and average cation exchange capacities (CEC) of selected soil colloids

Colloid	Source of Charge	Location of Charge	CEC Range ($cmol_c\ kg^{-1}$)	Average CEC ($cmol_c\ kg^{-1}$)
Fe and Al-oxides	pH dependent	surface	2-5	3
kaolinite	pH dependent	surface	2-10	5
Illite	Isomorphic substitution	tetrahedral	15-40	30
Montmorillonite	Isomorphic substitution	octahedral	80-120	100
allophane	pH dependent	surface	50-150	75
Humus	pH dependent	surface	100-225	150

Soil colloids most important to exchange reactions are the layer silicates, iron oxides, aluminum oxides, amorphous clays and humus. Each of these colloids has various sources of negative charge and different quantities of exchange sites present. There are two primary causes of negative charge in soils, **isomorphic substitution** and **dissociation reactions (H^+)** at the edge of the colloid. Isomorphic substitution is the substitution of an ion of similar size for another in during crystal formation. In crystal formation, the substituting ion often has fewer positive charges (e.g. Al^{3+} for Si^{4+}), thereby creating a permanent negative charge in the colloid crystal.

Dissociation reactions are caused by pH changes in the soil. Most minerals in the soil are composed of aluminum and silicon oxides. At the edge of these minerals are exposed O atoms containing a net negative charge. At low pH values, this charge is usually satisfied with a H^+ ion. However, as the pH increases, there are fewer H^+ ions in solution, so the charge is balanced by other cations present in solution. This source of negative charge is common for iron and aluminum oxides, 1:1 clays, and humic colloids. For all dissociation reactions, the CEC of the colloid is dependent on the pH of the soil. As pH increases, more H^+ leave the surface O atoms thereby increasing the total CEC.

Exchange Reactions

Cation exchange reactions involve one cation replacing another cation at the colloid surface. These reactions are controlled by the chemical and physical properties of the colloids and the ions in solution. The primary chemical property influencing the exchange reactions is the strength of electrical charge on the ions. The greater the ion valance, the stronger the electrical charge, and the stronger the attraction to exchange sites. The order in which ions are attracted to exchange sites is illustrated with the **lyotropic series**.

$$Al^{3+} > H^+ > Ca^{2+} > Mg^{2+} > K^+ = NH_4^+ > Na^+$$

As stated in most instances, the greater the valence the greater the ion attraction to the exchange sites, The exception is H^+ ions. Because H^+ ions are very small and are highly attracted to negative charges, they are more attracted to exchange sites than ions of higher valences.

The actual exchange of the ions on and of the exchange sites is a simple physical reaction where one ion replaces another. Therefore, if the concentration of one ion in soil solution is very high, it will overwhelm other ions currently on the exchange sites, regardless of the number of positive charges, due to the sheer number of ions present. The principle of overwhelming exchange sites with a single ion is used to measure the cation exchange capacity.

Cation Exchange Capacity

Cation exchange capacity (CEC) is a measure of the total number of positive charges needed to balance the negative charges per unit mass of colloid or soil (Eq. 10.1). The law of electoneutrality states that for every negative charge there must be a corresponding positive charge. What makes the CEC of a soil important for plant production is many of the cations that balance the negative charge on the colloid surface are often required for plant growth. Therefore, as the number of exchange sites increase, so does the potential number of plant nutrients. For this reason, it is important to know how many total exchange sites are present in a given soil. For the previously mentioned colloids, average CEC values are listed in Table 10.1. However, knowing the CEC of an individual colloid does not give a good picture of the CEC in soils because soils are a heterogeneous mixture of different colloids.

$$CEC = \frac{\text{number of sites}}{\text{wt. of colloid}}\left(\frac{\text{cmol}_c}{\text{kg}}\right) \qquad \textbf{Eq. [10.1]}$$

The CEC of a soil can be estimated using several different techniques. An estimate can be madeusing the type and amount of clay and organic matter in a soil and the average CEC values for those colloids. Another method is to measure the exchangeable cations present, assuming each charge associated with the cations represents an equivalent number of negative charges in the soil. Cation exchange capacity can also be estimated using cationic dyes. To do this, a series of soils with a range of known CEC values are treated with a cationic dye. The unknown samples are treated with the same dye and the color of the unknown sample leachate is then compared to the known CEC leachates to estimate CEC. A final method to determine CEC is to measure specific exchange reactions within the soil. To accomplish this, a solution containing one cation (e.g. K^+) is added to the soil to remove all the other cations from the colloid surfaces. A second solution containing a different cation (e.g. Ba^{2+}) is added to the soil. Where the second ion (Ba^{2+}) replaces all of the first ion (K^+) on the exchange sites and first ion (K^+) is collected and measured. The number of charges associated with the first cation (K^+) represents the number of exchange sites in the soil.

REVIEW QUESTIONS

1. Which soil colloid can be easily added to increase the CEC of a soil?
2. How does the type of soil colloid affect CEC?
3. What is the source of CEC for the different soil colloids?
4. What information does percent base saturation supply regarding nutrient status of a soil?
5. How does the calculation of surface area in Exercise 6 relate to CEC?

REFERENCES

Guertal, W.R., and P.A. McDaniel. 1990. A simple procedure for estimation of cation exchange using Gentian Violet dye. J. Agron. Educ. 19:189-190.

Jacobs, H.S. and R.M. Reed. 1964. Soils Laboratory Exercise Source Book. American Society of Agronomy. Madison, WI.

PROCEDURE

A. Charge on Soil Colloids

1. Place 2 funnels in a rack and label.
2. Fit each with filter paper.
3. Place the neck of each funnel in a test tube.
4. Add 5 g of soil to each filter paper.
5. Create a small depression in the soil surface.
6. Add Gentian Violet Dye (**GVD**) drop-by-drop to one of the soils until the soil is saturated. **Do not** let the dye run down the edge of the funnel. Make sure it passes through the soil. **[BE CAREFUL WITH THIS DYE. IT STAINS FOR A LONG TIME!]**
7. Stop adding the dye as soon as 5 drops of leachate are collected in the test tube.
8. Repeat steps 4 through 7 with the Eosin Red dye and the second soil.
9. Record your information in Table 10.2.

B. Cation Exchange Capacity

1. Select two soils from those available.
2. Weigh out 5 g of each into a plastic cup.
3. Add 50 mL of GVD to the soil.
4. Stir intermittingly with a glass rod for 10 min.
5. Between stirring periods, setup two funnels in a rack with filter paper (Whatman #2).
6. After 10 min., carefully pour the mixture into the funnel and collect the filtrate in a beaker. **Make sure that none of the mixture is allowed to run down the outside of the filter paper.**
7. Compare the color of your filtrate to standards prepared by the lab instructor to estimate the CEC of your soil.
8. Record your results in Table 10.3.

EXERCISE 11

Soil pH and Liming Acid Soils

OBJECTIVES

1. ***Know the definition of pH and how to calculate it***
2. ***Understand the difference between active, exchangeable, and reserve acidity***
3. ***Measure pH of soil using indicator dye and a pH meter***
4. ***Compare results between the methods***
5. ***Map results of pH from samples collected in the field from Exercise 3***
6. ***Determine the lime requirement of various soils***
7. ***Determine the effectiveness of liming materials***
8. ***Determine the amount of a specific liming material needed to meet the lime requirement of a soil***

INTRODUCTION

Soil pH

The measurement of pH often provides useful information on the status on many systems, of which soils are no exception. You are probably familiar with the terms acidic, basic (alkaline), and neutral. These are terms that refer to the relative amounts of hydrogen (H^+) and hydroxyl (OH^-) ions present in solution (Figure 11.1).

Relative Proportion	$H^+ > OH^-$	$H^+ = OH^-$	$H^+ < OH^-$
common term	acidic	neutral	basic (alkaline)
$[H^+]$	1	7.0	14.0
pH range	$1x10^{-1}$	$1x10^{-7}$	$1x10^{-14}$

Figure 11.1. The relationship between acidic ion concentration $[H^+]$, pH ranges and terminology

The pH is defined as the negative log of the hydrogen ion concentration $[H^+]$ (Eq. 11.1)

$$pH = -\log\ [H^+] = \log\frac{1}{[H^+]} \qquad \textbf{Eq. [11.1]}$$

where $[H^+]$ can be measured by several methods. Two of the most commonly used methods are:

1. Indicator dyes that are sensitive to $[H^+]$ (See Table 11.1).
2. Direct measurement of $[H^+]$ with H^+ sensitive electrodes and a voltmeter.

Of these methods, the pH electrode is the most accurate. However, it is not always convenient to send a sample to the laboratory for analysis, so if a field estimate is needed one can use a pH sensitive dye to get a quick picture of the soil pH. It is not recommended to use indicator dyes as anything other than a rough estimate of the soil pH. Dyes should **not** be used to determine application rates of lime or other nutrients to the soil. More accurate laboratory methods are needed to make amendment recommendations.

Table 11.1. Ranges of pH sensitivity for selected indicator dyes

Indicator	pH Range
Bromocresol Green	3.8 to 5.4
Chlorophenol Red	5.2 to 6.8
Bromthymol Blue	6.0 to 7.6
Phenol Red	6.8 to 8.4

Soil pH is affected by several factors including:

1. Type of clay mineral present
2. Number of cations present
3. Type of cations present
4. Concentration of salts in solution
5. Cations removed by plant uptake
6. Acid rain
7. Ammoniacal fertilizer additions
8. Leaching
9. Rainwater
10. Salt accumulation

Soil pH, in turn, affects many other soil properties including: cation exchange capacity, heavy metal availability microbial growth, and plant root growth. All plants have a pH range for optimal growth. Most plants prefer a pH value between 5.5 and 7.0, although there are plants that prefer pH values above and below this range. Low soil pH is often the most limiting factor for plant growth after plant available N, P and K deficiencies. Because many soil contaminates are more soluble in acid environments, soil pH is important in determining the potential environmental concerns of pollutants.

Soil acidity can be categorized as **active, exchangeable** or **reserve**. Active acidity is caused by H^+ ions in soil solution and is can be measured with soil pH. The source of exchangeable acidity is H^+ and Al^{3+} found on the cation exchange sites that can be easily removed by a replacement cation (neutral salt solution). Reserve acidity is a result of nonexchangeable aluminum hydroxides, aluminum and hydrogen. Reserve acidity is a long-term source of H^+ in the soil. The amount of active acidity in a soil is small when compared to the amount of exchangeable acidity. And both active and exchangeable acidity are small when compared to a soil's reserve acidity, as only a portion of the total amount of H^+ in the soil is in solution or on the exchange sites at a given period of time. Due to the importance of reserve acidity on soil pH (the greater the reserve acidity, the greater the $[H^+]$ in solution and the lower the pH), the number of exchange sites, the type of cations present and the percent base saturation will greatly affect soil pH. As previously mentioned, management of soil acidity can have a positive impact plant production when soil pH is maintained with in the optimal range for plant growth. Liming an acid soil is an effective way to decrease soil acidity.

Liming of Acid Soils

Soil pH will range from less than 4.0 in humid regions to greater than 8.0 in arid regions. In humid regions, increased rainfall leads to increased leaching of basic cations thus increasing concentrations of H^+ and Al^{3+} ions. Often highly weathered soils in humid regions are very acidic and suffer from significant pH-induced negative effects on biology (plant and microbial) and chemistry in the soil. When a soil pH falls below 5.0, it is difficult for most plant roots to thrive due to the presence of H^+ and Al-compounds. In addition, as pH decreases the exchangeable and reserve acidity of a soil increases and the basic cations required for plant nutrition decrease. To decrease soil acidity (i.e. raise the pH of a soil), liming materials are added to the soil. The composition of these materials varies depending on the region of the country, but regardless of composition, all liming materials contain basic cations and react within the soil to neutralize acidity. There are five steps in determining how much liming material to apply to a soil:

1. Determine recommended pH for the plant
2. Determine the current soil pH
3. Determine the amount of pure lime required to achieve the desired pH
4. Calculate the effectiveness of the available lime and the amount needed to reach the desired pH
5. Consider the cost effectiveness of the available lime

Liming Materials

Liming materials are used to adjust the pH of a soil and to maintain it at an optimum range for plant growth (OSU Fact Sheet 2212). A liming material must consume H^+ in solution **and** contain basic cations (most often Ca^{2+} and Mg^{2+}) (Table 11.4). Some materials, such as gypsum ($CaSO_4$), contain basic cations but do not neutralize H^+. Therefore, gypsum cannot be considered a liming material. The most common form of lime is $CaCO_3$, which undergoes the following reactions when added to the soil (Rxn 11.1, 11.2).

$$CaCO_3 + CO_2 + H_2O \leftrightarrow Ca(HCO_3)_2 \leftrightarrow Ca^{2+} + H_2O + CO_2(g) \qquad \textbf{Rxn. [11.1]}$$

Or

$$\boxed{\textbf{Colloid}}\begin{matrix}-H^+\\-Al^{3+}\end{matrix} + 2CaCO_3 + H_2O \longrightarrow \boxed{\textbf{Colloid}}\begin{matrix}-Ca^{2+}\\-Ca^{2+}\end{matrix} + 2CO_{2(g)}\uparrow + Al(OH)_3 \qquad \textbf{Rxn. [11.2]}$$

The Ca^{2+} released when $CaCO_3$ dissolves is available for exchange at the colloid surfaces. The Ca^{2+} will replace H^+ or Al^{3+} on the exchange site, thus decreasing the exchangeable acidity (Rxn. 11.2). Once in the soil solution, the lime neutralizes the H^+ and soil pH is increased. Since $CaCO_3$ is the most common form of lime used, all lime applications (tons $acre^{-1}$ or lbs $1000ft^{-2}$) are based on the equivalent weight of $CaCO_3$. Because other liming materials have different equivalent weights than $CaCO_3$, different quantities of these materials are required to achieve the same change in soil pH. In order to set these various liming materials relative to $CaCO_3$, the **calcium carbonate equivalent (CCE)** is used:

50

$$CCE = \left(\frac{\text{Eq. wt. } CaCO_3}{\text{Eq. wt. liming material}}\right) * \%\text{ purity} \qquad \textbf{Eq. [11.2]}$$

where:

CCE = Calcium carbonate equivalent (%),
Eq. wt. $CaCO_3$ = Molecular wt $CaCO_3$/ cation valence (g $mol_{(CaCO3}{}^{-1}{}_{)}$),
Eq. wt. material = Molecular wt liming material/cation valence (g $mol_{(x)}{}^{-1}$),
% Purity = Percent purity of the liming material.

When a liming material is added to the soil, some of the material will react quickly and the rest will react slowly over time. The time required for the material to react in the soil is a function of its purity and particle size. The size of the sorted particles is a major factor in the lime's reaction rate as it determines the amount of surface area exposed to the soil and soil solution. The **fineness factor (FF)** of the material is used to account for differences in size between liming materials and in Oklahoma is calculated using Eq. 11.3:

$$FF = \left(\frac{8\text{ mesh} + 60\text{ mesh}}{2}\right) \qquad \textbf{Eq. [11.3]}$$

where:

FF = Fineness factor (%)
8 mesh = percent passing through an 8 mesh screen with openings of 2.38 mm.
60 mesh = percent passing through a 60 mesh screen with openings of 0.25 mm.

All recommended liming rates are based on pure, finely ground $CaCO_3$, however it is pure, finely ground $CaCO_3$ is not used in agricultural applications due to expense. Therefore, application rates of other liming materials must be adjusted to an equivalent amount of calcium carbonate. This is the **effective calcium carbonate equivalent** of the liming **material ($ECCE_M$)** which is calculated by

$$ECCE_M(\%) = \left(\frac{CCE * FF}{100}\right) \qquad \textbf{Eq. [11.4]}$$

For example, if a liming material has an $ECCE_M$ of 65%, then 3,077 lbs. of this material will neutralize the same amount of acidity as 2,000 lbs. of pure, finely ground $CaCO_3$.

Lime Requirements of Soils

The **effective calcium carbonate equivalent required ($ECCE_R$)** is the amount of pure calcium carbonate liming material needed to increase the soil pH to the desired level. There are several ways to determine the $ECCE_R$ including: estimating lime need based on the amount and type of colloids present in the soil, precise titration measurements of the active exchangeable and reserve acidity, and empirical methods used in routine soil analysis.

Titration Method

Finding the $ECCE_R$ of a soil can be done by titrating a known mass of soil with a standard base ($Ca(OH)_2$) and calculating the milliequivalents of base added to raise the pH to the desired level. The number of milliequivalents of base added is a direct proportion of the amount of H^+ neutralized by the base and can be expanded to a field basis by calculating the number of milliequivalents needed on a hectare furrow-slice. Titration of each sample and back calculation to find the $ECCE_R$ is a time consuming., Therefore soil testing labs have developed more rapid methods to determine the exchangeable and reserve acidity of a soil. A common method is the **SMP buffer method.**

SMP Buffer Method

For the SMP buffer method the titration is done in advance. To determine the $ECCE_R$, a buffer solution at pH 7.5 is allowed to react with a known amount of soil. The pH of the buffer-soil mixture will decrease in proportion with the amount of reserve and exchangeable acidity present in the soil. The greater the decrease in the SMP buffer pH, the more lime is required to neutralize the soil acidity. Therefore, by measuring the final pH of the buffer-soil mixture, the $ECCE_R$ can quickly be determined using a table relating SMP buffer pH to tons of lime needed per acre to reach a desired field pH (OSU Fact Sheet 2212, Table 3).

need to know

1.) Find pH for plant (fact sheet) p. 95
2.) Determine pH of soil
3.) Determine $ECCE_R$ → Effective Calcium Carbonate Equivalent Required
4.) Determine $ECCE_m$ → m - material
5.) Cost Effectiveness

Calculation of Actual Lime Applied

Once the effectiveness of various liming materials ($ECCE_M$) and the amount of lime required to reach a desired pH ($ECCE_R$), are known, the amount of a specific liming material needed can be calculated using Eq. 11.5.

$$\text{Lime Applied}\left(\frac{\text{tons}}{\text{A}}\right)=\left(\frac{ECCE_R}{ECCE_M}\right)*100 \qquad \textbf{Eq. [11.5]}$$

With this information and total area of the area to be limed, the total amount of lime needed for a yard, garden, reclamation site, or field can be calculated.

REVIEW QUESTIONS

1. What is pH?
2. How does CEC effect pH?
3. How does pH affect plant growth?
4. Which pool of acidity is measured with a pH electrode?
5. Why do liming materials need to be put on an equivalent basis?
6. What are the requirements for a liming material?
7. What does $ECCE_M$ tell you about a liming material?
8. What does $ECCE_R$ tell you about a soil?
9. How does liming effect plant growth?

REFERENCES

Beyrouty, C.A., N. Slayton and M.G. Hanson. 1992. Lab Manual for Soils. University of Arkansas, Fayetteville, AR.

Jacobs, H.S. and R.M. Reed. 1964. Soils Laboratory Exercise Source Book. American Society of Agronomy. Madison, WI.

Shoemaker, H.E., E.O. McLean and P.G. Pratt. 1962. Buffer methods for determination of lime requirements of soils with appreciable amount of exchangeable aluminum. Soil Sci. Soc. Am. Proc. 25:247-277.

Yuan, T.L., 1974. A double-buffer method for the determination of lime requirements of acid soils. Soil Sci. Soc. Am. Proc. 38:437-440.

PROCEDURE

I. **Measuring Soil pH.**

A. **pH Meter**

1. Select two soils; an unknown supplied by the instructor and your sample collected from the field in Exercise 3.
2. Weigh out 10 g of each soil into separate plastic cups.
3. Add 20 mL of deionized water to each, stir with a glass rod, and then let sit for at least 20 min.
4. After 20 min. give the samples to the lab instructor to read the pH with the meter.
5. Record the results in Table 11.3.
6. Save both samples for part III.

B. **Indicator Dye.**

1. Use the same soils as in part A.
2. Obtain a spot plate for each and fill the large well approximately 2/3 full with soil.
3. Add 5 drops of deionized water so the soil is moist but not saturated.
4. Wait **three minutes** then add the indicator dye
   a) **Drop-by-drop** add the dye to the soil in the spot plate to the point of saturation.
   b) Once saturated add two more drops that will be used to determine the pH. **Do not add too much dye!**
5. Carefully swirl the soil-dye mixture to promote soil dye contact. **Do not stir the soil-dye mixture!!** Wait 3 min.
6. Tip the spot plate so that a drop of dye runs into the small well of the spot plate. A glass stirring rod may be used to lead the dye into the small well.
7. Determine the pH for your soil by comparing the color in the spot well with the standard card.

C. **Class pH**

1. In Figure 11.2 plot the pH at each location for the entire lab section.

II. **Preferred pH Ranges of Plants**

A. Complete Table 11.4. of the hand-in using OSU Extension fact sheet 2212.

III. **Determining Lime Requirement ($ECCE_R$) by SMP Buffer Method**

A. To the sample from I.A., add 10 mL of SMP buffer solution and stir intermittingly with a glass rod for 3 min.

B. Let the soil-buffer mixture set for 10 min. After 10 minutes, ask the lab instructor read the pH of the mixture using the pH meter.

C. Use Table 11.2 to determine the $ECCE_R$ for the soils and record the results in Table 11.6.

Lime Recommendations for Oklahoma

OSU Extension Facts

Published by Oklahoma State University   Distributed Through County Extension Offices   No. 2212

J. M. Baker
Extension Soil Specialist

A detailed discussion of the important causes and problems associated with treating soil acidity are discussed in Fact Sheet No. 2211. This Fact Sheet deals only with lime recommendations. It is essential that some terms be defined before progressing further.

Definition of Terms

AGRICULTURAL LIMING MATERIALS — Compounds capable of neutralizing soil acidity when added to the soil.

CALCIUM CARBONATE EQUIVALENT (CCE) — An expression of the acid neutralizing value of a material in comparison to pure calcium carbonate which has a value of 100%. This is determined by laboratory analysis.

FINENESS FACTOR — A value which describes the effectiveness of a liming material based on its fineness of grind. The fineness factor of materials passing through a 60 mesh sieve or finer is 100%. Coarser materials will have a lower fineness factor. The fineness factor is calculated as follows:

Fineness factor = One-half the percent material passing through a number 8 sieve plus one-half the percent material passing through a number 60 sieve. See example.

EFFECTIVE CALCIUM CARBONATE EQUIVALENT (ECCE) — the effectiveness of the lime material as it is affected by both purity (CCE) and fineness of grind. It is determined by multiplying the percent CCE by the fineness factor. See example.

Example: Determination of fineness factor and percent ECCE

	Neutralizing Value	*Fineness*	
Product	Percent CCE	Percent Passing 8 Mesh	Percent Passing 60 Mesh
Ground Limestone	90	80	60

The "Fineness Factor" is — 70 (½ of 80 + ½ of 60 = 70)
The Percent ECCE is — 63% (90 X 70 = 63%)

Limestone Materials

Liming materials are frequently used that have neutralizing values different from the value of pure calcium carbonate. The average neutralizing values ($CaCO_3$ equivalent) and the chemical formulas of some important sources are given in Table 1. This means for example, that 100 pounds of burned lime may be equivalent to as much as 150 pounds of pure calcium carbonate.

Table 1. Basic Soil Conditioning Materials.

Name	*Chemical Compound*	*Avg. Percent $CaCO_3$ Equivalent*
Shell marl	$CaCO_3$	95
Agricultural limestone	$CaCO_3$	95
Hydrated lime	$Ca(OH)_2$	120
Burned lime or builders lime	CaO	120 to 150
Dolomite	$CaCO_3 . MgCO_3$	110
Marl	$CaCO_3 . H_2O$	15 to 85

Fineness

The effectiveness of any limestone material is related to the fineness of the individual particles. Generally, the smaller the particle size the more effective the material. Also, finer materials react in the soil much more rapid which results in a quicker adjustment of soil pH. This occurs because the surface area of fine particles is much greater than the surface area of coarse particles. Consequently, the finer particles dissolve to a much higher degree and much quicker than coarse particles. In addition, uniform application of fine particles is accomplished much easier than coarse particles.

Application

Generally it is desirable to neutralize the surface 5 to 6 inches of soil. Obviously this is accomplished best if the lime material is mixed into this zone after application by disking. While it is recognized that this is the best way, it is not always possible. For instance, lime cannot be incorporated on established stands of alfalfa or on permanent pasture without resulting damage to the crops. In these cases it is acceptable to broadcast the material on the surface, but it should be remembered that effectiveness will decrease somewhat and time of reaction will increase. Nevertheless it is much better to apply the material in this manner than not apply it.

Rate of Application

The rate of limestone material that should be applied to a soil is dependent on several factors. They are:

1. Quality of limestone material (ECCE)
2. Present pH of the soil
3. Desirable pH for the crop to be grown
4. Resistance of the soil to change (buffer capacity)

QUALITY: Obviously more limestone material is required to adjust pH when the material is of low quality than if it is of high quality. Therefore, before knowing how much limestone material to apply one must know the % ECCE of the available material. Legislation enacted July 1, 1973 requires that the % ECCE be given to the buyer on all agricultural limestone sales. Since that time, pounds of ECCE are recommended to farmers rather than pounds of lime material. The limestone material requirement can be obtained from Table 2 or it can be calculated as follows:

$$\text{Limestone Material Requirement} = \frac{\text{lbs of ECCE Recommended}}{\text{\% ECCE}} \times 100$$

PRESENT pH OF THE SOIL: When the pH of the soil is below the suggested range for the crop to be grown, liming should definitely be considered. It is generally believed that anytime the soil has a pH of less than 5.5 an adjustment should be made. It is at this point that secondary problems such as toxicity from aluminum and manganese start occurring.

DESIRABLE pH OF THE SOIL: It has been general practice to lime soils to a pH of 6.8, although this is not always necessary. The pH of the soil should be adjusted to within the suggested range for the crop to be grown.

RESISTANCE OF THE SOIL TO CHANGE: Soils differ widely in their buffer capacity which is their resistance to pH changes. Two to three times as much ECCE will be required to adjust the pH of a soil with a high buffer capacity as is required to adjust the pH of a soil with low buffer capacity. Generally sandy soils will have a lower buffer capacity than finer textured soils. This property of soils is discussed in detail in Fact Sheet 2211. Before determining the ECCE requirement of a soil, some indication of the buffer capacity of the soil is needed. The OSU Soil and Water Laboratory conducts a test on all soils with a pH of 6.4 or less called the Buffer Index. This test makes it possible to account for the buffer capacity of the soil in arriving at the proper rate of ECCE to be applied. Experience has shown that most of the soils in Oklahoma have a low buffer capacity, i.e. the pH of the soil can be adjusted easily. This is reflected by the fact that the buffer index of most soils is 6.8 which gives an ECCE requirement of 1.2 tons per acre (Table 3).

How to Determine ECCE Requirements

1. A buffer index will be determined on soils having a pH of 6.4 or less.

2. Refer to Table 3 for the lime requirement for each buffer index.

3. If the pH of the soil is less than 6.2, a minimum of 1 ton of ECCE per acre should be applied to alfalfa regardless of the buffer index. If the buffer index indicates need for more than 1 ton ECCE, the higher rate should be applied.

4. Usually due to application cost it is not feasible to apply less than 1 ton ECCE per acre.

5. When the buffer index indicates a need for more than 3 tons ECCE per acre, the application should be split to facilitate spreading and mixing with the soil. No more than 2 tons ECCE per acre should be applied to established alfalfa or pastures at any one time.

Table 2. Pounds of Liming Material Required for Different Rates of Effective Calcium Carbonate Equivalent (ECCE) and Percent ECCE.

Percent	Rate of ECCE Recommended in Pounds									
ECCE	500	1,000	1,500	2,000	2,500	3,000	3,500	4,000	4,500	5,000
30	1,650	3,350	5,000	6,600	8,350	10,000	11,500	13,350	15,000	16,650
40	1,250	2,500	3,750	5,000	6,250	7,500	8,750	10,000	11,250	12,500
50	1,000	2,000	3,000	4,000	5,000	6,000	7,000	8,000	9,000	10,000
60	800	1,650	2,500	3,350	4,150	5,000	5,850	6,650	7,500	8,350
70	700	1,450	2,150	2,850	3,550	4,300	5,000	5,700	6,450	7,150
80	600	1,250	1,900	2,500	3,100	3,750	4,400	5,000	5,650	6,250
90	550	1,100	1,650	2,200	2,800	3,350	3,900	4,450	5,000	5,550
100	500	1,000	1,500	2,000	2,500	3,000	3,500	4,000	4,500	5,000
110	450	900	1,350	1,800	2,250	2,750	3,200	3,650	4,100	4,550
120	400	850	1,250	1,650	2,100	2,500	2,900	3,350	3,750	4,150
130	400	750	1,150	1,550	1,900	2,300	2,700	3,100	3,450	3,850
140	350	700	1,050	1,450	1,800	2,150	2,500	2,900	3,200	3,550

Table 3. ECCE Required to Bring Soil to an Indicated pH According to Soil-buffer Index (Neutralization of 6-7 Inch Furrow Slice).

Soil-buffer index	ECCE* reqd. to bring soil to indicated pH(tons/A)		
	7.2	6.8	6.4
Over 7.1	.5	none	none
7.1	.7	.7	none
7.0	1.0	.7	none
6.9	1.2	1.0	none
6.8	1.4	1.2	.7
6.7	1.6	1.4	1.2
6.6	2.1	1.9	1.7
6.5	2.8	2.5	2.2
6.4	3.5	3.1	2.7
6.3	4.2	3.7	3.2
6.2	4.7	4.2	3.7
6.1	5.2	4.8	4.2
6.0	5.9	5.4	4.7

* Pure $CaCO_3$ ground enough to be 100% effective.

RELATIVE VALUE OF LIME MATERIALS: Since agricultural lime materials vary considerably in quality they also vary in their relative dollar value. For instance, material that is 80% ECCE would be worth twice as much as a material with an ECCE of only 40%. Table 4 gives the relative value of lime materials. In the above case, it indicates that if 40% ECCE material cost $3.20/ton then one could afford to pay as much as $6.40/ton for 80% ECCE. Another example would be if 60% ECCE material cost $7.20/ton, then 50% ECCE material would have a value of only $6.00 per ton.

Table 4. Relative Value of Limestone Materials Based on their % ECCE

Percent ECCE									
30	40	50	60	70	80	90	100	110	120
Relative Value/ton									
2.40	3.20	4.00	4.80	5.60	6.40	7.20	8.00	8.80	9.60
3.00	4.00	5.00	6.00	7.00	8.00	9.00	10.00	11.00	12.00
3.60	4.80	6.00	7.20	8.40	9.60	10.80	12.00	13.20	14.40
4.20	5.60	7.00	8.40	9.80	11.20	12.60	14.00	15.40	16.80
4.80	6.40	8.00	9.60	11.20	12.80	14.40	16.00	17.60	19.20
5.40	7.20	9.00	10.80	12.60	14.40	16.20	18.00	19.80	21.60

Table 5. pH Preference of Plants

Farm Crops and Grains

	Suggested pH Range	pH Tolerance
LEGUMES		
Alfalfa	6.5-7.5	6.0-8.0
Alsike Clover	6.0-7.0	5.0-8.0
Cowpeas	5.5-7.0	5.0-8.0
Crimson Clover	5.5-7.0	5.0-8.0
Red Clover	6.0-7.0	5.5-8.0
Rye	5.5-7.0	4.5-7.5
Soybeans	5.5-7.0	5.0-8.0
Sweet Clover	6.5-7.5	6.0-9.0
Vetch	5.5-6.75	5.0-7.5
White Clover (Ladino)	6.0-7.0	5.5-8.0
NON-LEGUMES		
Barley	6.5-7.0	5.5-9.0
Buckwheat	5.3-6.5	4.0-7.0
Corn	5.5-7.0	5.0-7.5
Cotton	5.5-6.75	4.5-8.5
Oats	5.5-7.0	5.0-7.5
Sorghum	5.5-7.0	5.0-8.5
Wheat	5.5-7.0	5.0-8.5
GRASSES		
Bermuda	6.0-7.0	5.0-9.0
Fesque	4.5-7.0	4.0-8.0
Sudan Grass	5.5-7.0	5.0-8.5

Vegetable Crops

	Suggested pH Range	pH Tolerance
VEGETABLES		
Asparagus	5.5-7.0	5.0-7.8
Beans	5.5-6.5	5.0-7.0
Peas	5.5-6.5	5.0-7.0
ROOT CROPS		
Beets	6.0-7.0	5.5-7.0
Carrots	5.5-6.5	5.0-7.0
Irish Potatoes	5.0-5.5	4.5-7.5
Onions	6.0-7.0	5.5-7.0
Parsnips	6.0-7.0	5.5-7.0
Radishes	5.5-6.5	5.0-7.0
Sweet potatoes	5.0-6.8	5.0-6.8
Turnips	5.5-6.5	5.0-7.0
LEAFY VEGETABLES		
Cabbage	6.0-7.0	5.5-7.0
Collards	6.0-7.0	5.5-7.0
Lettuce	6.0-7.0	5.5-7.0
Mustard	6.0-7.0	5.5-7.0
Turnip-greens	6.0-7.0	5.5-7.0
Spinach	6.0-7.0	5.5-7.0
VINE CROPS		
Cantaloupe	5.5-6.8	5.0-6.8
Cucumbers	5.5-6.8	5.0-6.8
Pumpkin	5.5-6.8	5.0-6.8
Squash	5.5-6.8	5.0-6.8
Watermelon	5.5-6.8	5.0-6.8
FRUITING VEGETABLES		
Corn	5.5-6.8	5.0-7.0
Eggplant	5.5-6.8	5.0-7.0
Okra	5.5-6.8	5.0-7.0
Pepper	5.5-6.8	5.0-7.0
Tomato	5.5-6.8	5.0-7.0

Fruits

	Suggested pH Range	pH Tolerance
BRAMBLE FRUITS		
Blackberries	5.0-6.5	5.0-8.0
Black Raspberries	5.0-6.5	5.0-8.0
Dewberries	4.5-5.5	4.0-6.5
Grapes	5.5-6.5	4.5-8.0
Red Raspberries	5.0-6.5	5.0-8.0
Strawberries	5.0-6.5	5.0-8.0
TREE FRUITS		
Apples	5.0-6.5	4.5-7.5
Apricots	5.5-6.5	4.5-7.5
Cherries (sour)	5.5-6.5	4.5-7.2
Crab Apples	5.5-6.5	4.5-7.5
Peaches	5.5-6.5	4.5-7.2
Pears	5.5-6.5	4.5-7.5
Plums	6.0-7.0	4.5-7.8

Nut Trees

	Suggested pH Range	pH Tolerance
Pecan	5.5-6.5	5.0-7.0

Ornamentals

SHRUBS—DECIDIOUS	Suggested pH Range
Barberry	6.0-7.5
Cotoneaster	6.0-7.0
Crape Myrtle	5.5-7.5
Deutzia	6.0-7.5
Flowering Quince	6.0-7.0
Forsythia	6.0-7.0
Lilac	6.0-7.5
Privet	6.0-7.5
Spirea	5.5-7.0
Weigela	6.0-7.0

TREES—NARROW LEAF EVERGREEN

	Suggested pH Range
Arbovitae	6.0-7.5
Arizona Cypress	6.0-7.5
Japanese Yew	6.0-7.0
Mugho Pine	6.0-7.5
Prostrate Juniper	6.0-7.5
Red Cedar	6.0-7.5

TREE—BROAD LEAF EVERGREEN

	Suggested pH Range
Abella	6.0-7.5
Box	6.0-7.0
Euonymus	6.0-7.0
Holly	5.5-7.0
Liriope	6.0-7.5
Live Oak	6.0-7.5
Magnolia	5.5-6.5
Photinia	6.0-7.0
Pyracantha	6.0-7.5
Rhododendron	4.5-5.5
Yucca	6.5-7.5

TREE—DECIDUOUS

	Suggested pH Range
Ash	6.0-7.5
Birch	5.0-6.5
Bur Oak	6.0-7.5
Catalpa	6.0-8.0
Cottonwood	6.0-8.0
Elm	6.0-7.5
Flowering Plum	6.0-7.5
Golraintree	6.0-7.5
Mimosa	6.0-7.5
Pin Oak	6.0-7.0
Redbud	5.5-7.5
Sugar Maple	6.5-7.5
Sweetgum	6.0-7.0
Sycamore	6.0-7.5
Weeping Willow	5.5-7.0

$O \rightarrow 2$

$CO_3 \rightarrow 2$

$OH \rightarrow 1$

$H_2O \rightarrow 0$

Oklahoma State University extension programs serve people of all ages, regardless of socio-economic levels, race, color, sex, religion, or national origin. Issued in furtherance of Cooperative Extension work, acts of May 8 and June 30, 1914, in cooperation with the U.S. Department of Agriculture, J. C Evans, Vice President for Extension, Director of Cooperative Extension Service, Oklahoma State University, Stillwater, Oklahoma. 10-73/22¾M Rev. 05